The Naked Scientists

BOOM!

50 FANTASTIC SCIENCE EXPERIMENTS
TO TRY AT HOME WITH YOUR KIDS

Learn how to extract DNA from a kiwifruit (page 53).

The Naked Scientists

BOOM!

50 FANTASTIC SCIENCE EXPERIMENTS
TO TRY AT HOME WITH YOUR KIDS

Chris Smith & Dave Ansell—The Naked Scientists®

THE NAKED SCIENTISTS

IMM **lifestyle**
books™

Read. Learn. Do What You Love.

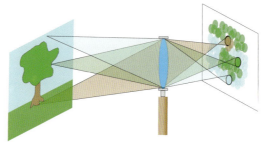

CONTENTS

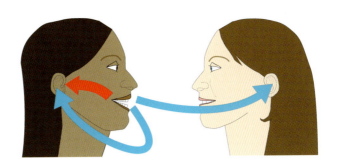

MAKE YOUR OWN SUBMARINE

In this experiment we'll find out how submarines sink and resurface and how scuba divers can hover in the water without bashing into the bottom. It's all about buoyancy and how you can control it.

◆ Take a large (2-quart/2-liter) plastic soft drink bottle and fill it with water.

◆ Add an unopened plastic packet of ketchup or mayonnaise (like the ones given out at fast-food restaurants). The packet will just about float, but if it's too buoyant try attaching one or more paperclips or modeling clay to the outside so that it sits just at the surface of the water.

◆ Make sure that the bottle is full to the brim.

◆ Now screw on the top very tightly and squeeze the bottle hard.

◆ The sauce submarine will dive to the bottom of the bottle.

WHY does it work?

It's all down to a piece of physics worked out by the famous Greek "eureka-streaker," Archimedes. When he leapt out of his bath 2,000 years ago, what Archimedes had effectively discovered was why things float. He found that water pushes up on an object with a force equal to the weight of the water that the object displaces or pushes out of the way. If the displaced water is heavier than the object itself, then it will float upward.

When the ketchup packet is first added to the plastic bottle, it displaces an amount of water that weighs slightly more than it does, and so it bobs around on the surface. When you squeeze the closed bottle, you apply pressure to the water inside. Liquids are incompressible, so the pressure is instead transmitted to the ketchup packet; this contains a small pocket of nitrogen gas (to keep the ketchup fresh).

Unlike liquids, gases can be compressed, so the squeezing effect of the water causes the packet to begin to shrink. This reduces the amount of space it takes up and, hence, the amount of water that it displaces, even though its weight isn't changing. Eventually, once you squeeze hard enough, the packet will shrink to a point where the amount of water being displaced weighs less than the ketchup packet does, and it will sink.

Why does it surface again when you release the pressure? This is because, as soon as you let go, the gas in the packet re-expands, making the packet less dense than the water around it, so it floats upward.

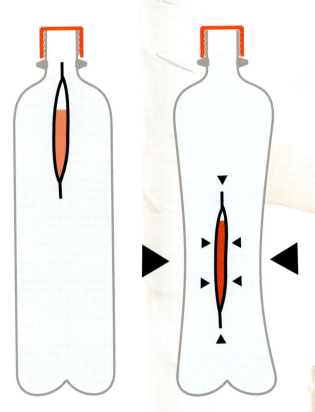

HOW does this apply to the real world?

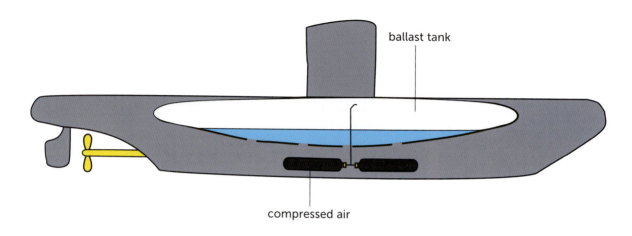

ballast tank

compressed air

Both scuba divers and submarines control their depth using similar principles. Scuba divers wear a BCD (buoyancy control device). This is an inflatable jacket that behaves just like the sauce packet in this experiment, but with the exception that extra air can be pumped into it from the cylinder on the diver's back. When the BCD inflates, it blows up like a balloon, displacing more water, so that the diver floats upward.

Subs are slightly different. They have ballast tanks around the outside of the vessel. When the submarine dives, some of the air in these tanks is replaced with water, increasing the weight of the vessel. This makes it heavier than the water it's displacing, so it sinks. To resurface, compressed air stored in tanks inside the submarine is blown into the ballast tanks where it pushes out the water. This makes the vessel weigh less than the displaced water, so it rises again.

SOME OTHER THINGS TO TRY

Try dissolving some salt in the water in your bottle. The more salt you dissolve, the harder it becomes to persuade your ketchup-packet submarine to sink. This is because the dissolved salt increases the weight and, therefore, the density of the water, making the packet more buoyant. This is why ships can carry more cargo at sea than they can in fresh water and why it's almost impossible to sink in the Dead Sea.

CREATE A CLOUD IN A BOTTLE

Clouds come in all shapes and sizes, but how do they form in the first place? In this experiment we'll make clouds appear and disappear, with just the squeeze of a bottle.

ADULT SUPERVISION!

1

2

◆ Take a large (2-quart/2-liter) plastic soft drink bottle and add a small amount of water.

◆ Light a match, blow it out, and drop it into the bottle while it's still smoking.

◆ Close the bottle tightly.

◆ Squeeze the bottle really hard for about 5–10 seconds, and at the same time give it a swirl.

◆ Stop squeezing and let the bottle expand again.

◆ You should see a cloud appear inside the bottle. Squeeze the bottle again, and it will disappear. Let it go, and it will return.

WHY does it work?

This experiment relies on the physics that powers both a diesel engine and a refrigerator. The sealed bottle is filled with air molecules and a small number of water molecules. When you squeeze the bottle, the energy you use is transferred to the gas molecules inside, causing them to heat up. You can demonstrate this another way by placing your thumb over the end of a bicycle pump as you push down the plunger—you'll feel your thumb getting hot.

This is exactly how a diesel engine works; the cylinders compress air, heating it up to several hundred degrees Farenheit (Celsius). The diesel is then injected and instantly ignites, releasing the energy it contains.

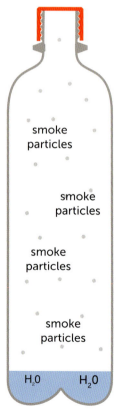

smoke particles

smoke particles

smoke particles

smoke particles

H_2O H_2O

The bottle contains smoke particles from the match.

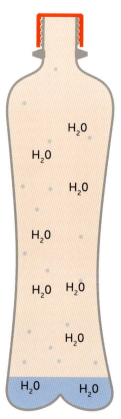

H_2O

H_2O

H_2O

H_2O

H_2O H_2O

H_2O

H_2O H_2O

Squeezing the bottle raises the temperature, evaporating some of the water.

H_2O H_2O

Let go! Water droplets condense on the smoke particles, forming a cloud.

In our experiment, the extra heat generated by squeezing the bottle encourages some of the water at the bottom of the bottle to evaporate, forming invisible water vapor.

Then, when you allow the bottle to expand again, the reverse effect kicks in. The sudden drop in pressure causes the temperature to fall. Cold air cannot hold as much water vapor as warm air, so the water molecules that evaporated before now begin to link up with each other to form small droplets of liquid water. This is the cloud that appears.

Where does the smoking match come in? It's very hard for water molecules to condense in clean air. Instead, they look for a surface where they can congregate, and the smoke particles drifting around inside the bottle provide a perfect place for this to happen.

HOW does this apply to the real world?

Most of the clouds you see in the sky form in a similar way. Heat from the Sun evaporates water from the Earth's surface, and this vapor is carried aloft by rising warm air. As it climbs further from the planet's surface, the atmospheric pressure drops, causing the air to expand and cool like it did when you released the bottle. Eventually, the air cools to a point where it can no longer hold the water vapor and the water vapor begins to condense.

Just as in our experiment, it does so on particles of dust, pollen, and pollution, that are also in the air. Together, the billions of tiny water droplets form a cloud. When the droplets become large enough, they fall as rain, hail, or snow.

What about fog, is that the same? Sometimes the temperature close to the ground can fall very fast, making water vapor condense at ground level. If there's no wind to blow it away, the result is mist and fog.

SOME OTHER THINGS TO TRY

Repeat the experiment without the smoke from the match. It's very difficult to achieve the same effect. Or try breathing on a mirror. You'll see it gets foggy. This is because the cold surface removes energy from the water vapor in your breath, so it condenses on the glass surface, forming water droplets.

11

LEVITATING PING PONG BALLS

An antigravity machine would certainly be an asset, and although we can't claim to have invented one, in this experiment we'll show you a feat of physics that keeps planes aloft and will also hold a ball suspended in midair.

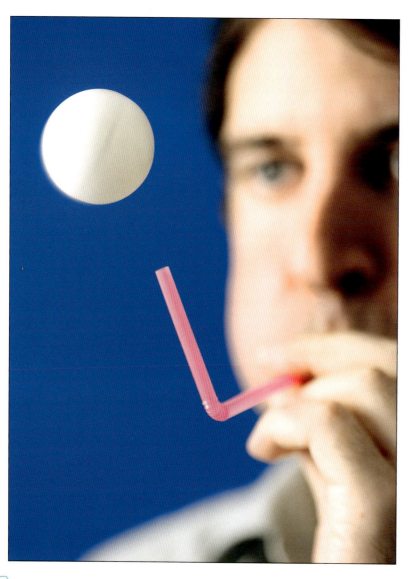

- ◆ For this experiment you need a ping pong ball and either a hairdryer or a bendy drinking straw.

- ◆ Turn on the hairdryer (set to cold if possible).

- ◆ Point it so it's blowing a stream of air straight upward. (If you're using a straw, put the long part in your mouth with the shorter bendy section pointing toward the ceiling and blow hard.)

- ◆ Hold the ping pong ball in the airstream coming from either the hairdryer or the straw, and then let it go. Mysteriously, it bobs about in the air flow without falling off.

- ◆ Try angling the airstream, and the ball will follow, even when the flow is tilted over 30 degrees.

WHY does it work?

The air leaving the straw blows the ping pong ball upward, but what keeps it there, and why doesn't it fly off and fall to the floor? The answer is an effect described in the 1930s by the Romanian airplane designer Henri Coanda. He showed that when air, or a fluid, flows over a curved surface, it can stick to the surface and follow it, so the flow also becomes curved. This means that when the ping pong ball sits in the center of the airstream, the flow traps the ball by passing around it on all sides, sticking to its surface.

Why does it remain in one place? This is thanks to Isaac Newton's Third Law, that states that for every action there must be an equal and opposite reaction. If the ball tries to move in any direction, the air sticking to its surface will be pulled with it. But if the air is being moved, then there must also be a force pushing back on the ball in the opposite direction, which is what holds it steady. This is why it bobs about in one place, even when the airstream is at an angle.

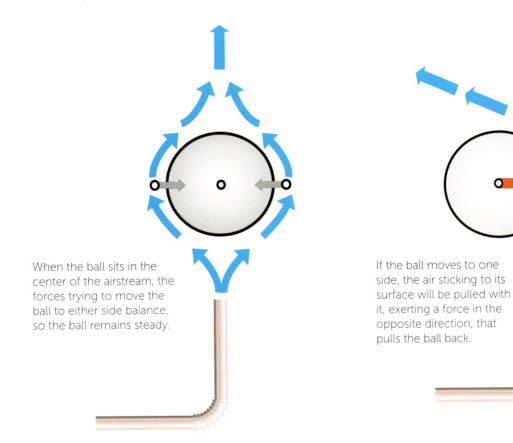

When the ball sits in the center of the airstream, the forces trying to move the ball to either side balance, so the ball remains steady.

If the ball moves to one side, the air sticking to its surface will be pulled with it, exerting a force in the opposite direction, that pulls the ball back.

HOW does this apply to the real world?

This phenomenon is also the means by which the wing of an airplane generates lift. Air passing below the wing is deflected downward by the curved surface. At the same time, air sticks to the top of the wing due to the Coanda Effect. Because the air is pushed downward by the wing, there is an equal and opposite upward force applied to the plane, keeping it airborne. This is the same principle that held the ball steady in the airstream in this experiment.

However, if the pilot tries to climb too quickly, the airflow can detach from the top surface of the wing, creating an area of swirling turbulence that does not generate any lift. This is called a stall and it can be very dangerous close to the ground because the loss of lift causes the plane to lose altitude very rapidly and, potentially, to crash.

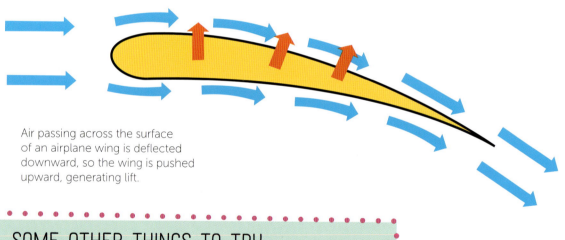

Air passing across the surface of an airplane wing is deflected downward, so the wing is pushed upward, generating lift.

SOME OTHER THINGS TO TRY

You can also demonstrate the Coanda Effect by trickling water from the tap down the back of a tablespoon. Rather than running straight down the back of the spoon, the stream curves around the tip, coming off at an angle.

RICE QUICKSAND

In this experiment, we'll show you how to pick up a jar of rice without touching it and why this can explain holes in the road...

◆ Take a jam jar and fill it with rice. Basmati rice works best.

◆ Take a knife the same length as the jar and push it all the way into the rice.

◆ Wiggle the knife about, then remove it and reinsert it.

◆ Keep doing this, adding more rice to the jar whenever the level falls.

◆ After several minutes, you will notice that it's becoming more and more difficult to insert the knife. Eventually, you will be able to lift up the jar just using the knife embedded in the rice.

WHY does it work?

It's all down to how particles organize themselves, and it's also the reason why patios and pavements end up uneven, why quicksand is so deadly, and why roads often become full of potholes.

When you first fill the jar, the grains take up random positions with large gaps between them. This makes it very easy to insert the knife because the rice can move out of the way and into the gaps.

As you repeat the process, the rice grains begin to pack together much more tightly and in a more organized fashion. There's now much less empty space, which is why you needed to top off the jar with rice along the way. If you look at the jar from the side, you'll see that most of the grains are lined up in rows.

In this compact arrangement, the rice takes up much less space; this increases the density of the rice so that it becomes

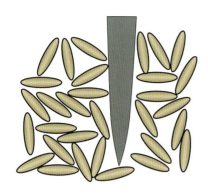 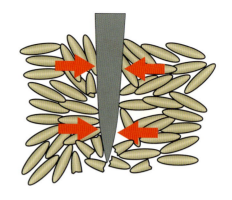

progressively more difficult for the grains to move out of the way of the blade. Eventually, it's only possible to insert the knife by breaking, cutting, or distorting the grains, which requires a large amount of force.

The result is that the grains push back on the knife with the same amount of force, creating an effective "rice-vice" that grips the blade firmly with more force than it takes to lift the jar.

HOW does this apply to the real world?

What's this got to do with patios, pavements, potholes, and quicksand?

When paving stones are first laid down, or when a hole in the road is repaired, the result is usually a flat surface. After a while, vibration from traffic, or people walking over the surface, causes the small particles in the repair or foundation material to compact together.

Just as the level of rice fell in the jar in this experiment, the level of material filling the hole or supporting the paving slab also drops, making a dip in the road or raising a corner of the stone that people can trip over.

And the quicksand? Well, this is a similar principle. Quicksand is composed of a mixture of salt water and sand particles glued together with small amounts of clay. It forms a structure rather like a house of cards, with large water-filled spaces between each of the sand particles.

When you tread on the quicksand, the pressure you apply breaks down the house of cards, and all of the particles pack tightly together around the trapped body part, locking it in place. In fact, the sand is so heavy that the force needed to pull you out is greater than you would need to lift a car.

Don't believe everything you see in the movies, though, because although you might become stuck, it's impossible to drown in quicksand: it's twice as dense as you are, so you'll only sink to waist height.

HOMEMADE LITMUS TEST

Chemicals that are either acids or bases (alkalis) can look very similar, so it can be hard to tell them apart. In order to do so, chemists use substances called indicators that change color accordingly. The most famous example is litmus, which is made from lichens, but other plants and vegetables also contain pigments that can work the same way.

YUCK! DON'T EAT!

1

2

◆ Take about a quarter of a red cabbage.

◆ Use a shredder or a knife to slice the cabbage into small pieces.

◆ Transfer these to a bowl and add a small amount of water (but not enough to cover the cabbage).

◆ Crush the cabbage with a wooden spoon, and then pour the result through a sieve, keeping just the liquid. It should be a purple-blue color. This is your indicator test solution.

◆ Add small amounts of this indicator solution to a series of plastic cups or glasses—you need at least three.

◆ Aim to keep one cup of indicator as your "control." Don't add anything to this cup so that you can use it as a record of the starting color of the solution.

◆ To one of your other cups, add a teaspoonful of lemon juice or vinegar, and to another cup a teaspoonful of baking soda (bicarbonate of soda).

◆ Swirl the cups to mix the contents, and then watch what happens. The liquid in the vinegar (or lemon juice) cup should turn pink, and the liquid in the baking soda cup should turn dark blue.

WHY does it work?

It's down to a discovery first made by the seventeenth-century chemist Robert Boyle, who coined the terms "acid" and "alkali" (or "base") and who also invented the litmus test.

Red cabbage contains a water-soluble pigment called a flavin, part of a family of chemicals called anthocyanins. These are large molecules consisting of several linked rings of atoms that share their electrons amongst themselves. These electrons absorb certain wavelengths of visible light but reflect others. The reflected light is what gives the chemical its characteristic color.

When an acid is added, such as the acetic acid in vinegar or the citric acid in lemon juice, the acid adds extra hydrogen (known as hydrogen ions) to some of the oxygen atoms in the anthocyanin molecule.

This prevents some of the electrons from being shared, causing them to soak up more blue light. Since white light is a mixture of wavelengths ranging from blue to red, removing more of the blue light has the effect of making the chemical look more red.

What about when an alkali is added, such as baking soda? Unlike acids, that try to give hydrogen ions to other chemicals, alkalis or bases try to remove them. When baking soda, that contains sodium bicarbonate, is added to the cabbage indicator, it steals some of the hydrogen from the molecule. To make up for the loss of the hydrogen, the anthocyanin shares out more of its electrons around the molecule. This causes it to absorb more red light and to reflect more blue light, so it looks blue.

HOW does this apply to the real world?

Most of the colors you see in nature are thanks to the reaction you've reproduced in this experiment. This was discovered by the German chemist and 1920 Nobel prizewinner Richard Willstätter. He showed that roses are red and violets are blue not because they make different pigments but because they alter the acidity or alkalinity of their petals. This, in turn, changes the color of their anthocyanins, altering the color of the flower.

SOME OTHER THINGS TO TRY

Use your indicator solution to test other household substances. Good things to try are soap, sour milk, cream of tartar, carbonated water, and a soft drink. Also, prove that the indicator effect is reversible. Add something acidic to your indicator, and then add something alkaline.

Try other plants from the garden. Which of them work as indicators? Is there anything that connects the ones that do?

IRON OUT YOUR CEREAL

"Fortified with iron" says the side of the cereal box. But is there really iron in there and, if so, what does it do, and can you see it?

- For this experiment, you will need some breakfast cereal, preferably one claiming to be high in iron, a mortar and pestle, or something similar with which to crush it, and a strong magnet or pair of magnets.

- Take a cupful of the cereal and grind it to a fine powder.

- Add one of the magnets and mix it round in the cereal dust.

- Retrieve the magnet and study the surface.

- You should be able to see fine grains of cereal sticking to it, and if you bring a second (stronger) magnet near to the first, the particles will jump onto the second magnet.

WHY does it work?

It works because, during manufacture, the cereal has been laced with particles of metallic iron that are magnetic. This is done to help you to stay healthy, because iron (its symbol is "Fe") is vital to your metabolism. It's known for its role in hemoglobin, the pigment in red blood cells that carries oxygen from the lungs to the tissues, but it's also an essential part of a family of enzymes called cytochromes. These help to generate energy in cells and also break down drugs and toxins. The body needs adequate supplies of iron in order to function efficiently.

Without sufficient iron, a person can become anemic, a condition caused by having too few red blood cells. This can lead to tiredness, irritability, poor concentration and, in severe cases, feeling faint and breathless because your body can't transport enough oxygen to your brain and tissues.

To reduce the chance of this happening, it's important to eat a healthy diet. The best source of iron is red meat, including beef, pork and liver. This source of iron is referred to as "haem-iron" and is in a form that can be most efficiently used by the body.

Some plant produce also contains reasonable amounts of iron, including baked beans, kidney beans, apricots, and spinach, but this is in a form known as "non-haem iron," that is more difficult for the body to use. However, vitamin C has recently been found to help the body to absorb and use this form of iron, so a glass of orange juice with meals is a good idea for vegetarians.

HOW does this apply to the real world?

Because some people may not take in enough iron in their diet and are therefore at risk of iron deficiency, additional iron is added to some foods, such as cereals. This is known as "fortification." Cereal makers have found that the best way to do this is by adding tiny particles of metallic iron, smaller than a human hair. These react with the acid in the stomach to produce iron chloride; this can be absorbed by the small intestine. From there, it's picked up by a protein called "transferrin," that shuttles the iron to where it's needed in the body. Nonmetallic forms of iron can cause food to spoil, which is why manufacturers avoid using them.

SOME OTHER THINGS TO TRY
Use the same technique to test a variety of cereals. Do they all contain detectable iron particles?

HURRICANE IN A BOTTLE

Hurricanes are spectacular circling storm systems where an area of low pressure about 20 miles (30 kilometers) across, known as the eye, is surrounded by a mass of swirling air. The largest storms are over 400 miles (650 kilometers) in diameter and can generate wind speeds exceeding 100 miles (160 kilometers) per hour. How do they work, and why is the "eye" calm and wind free?

- For this experiment, you will need two large plastic soft drink bottles, some water and some duct or insulating tape.
- Fill one of the bottles two-thirds full with water.

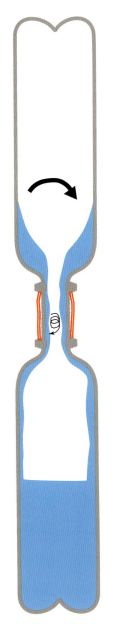

- Invert the second (empty) bottle over the first and tape the two necks together so that the whole thing resembles a large egg timer.
- Turn the bottles upside down so that the one containing the water is on top.
- Watch how the water makes its way from the upper to the lower bottle.

- When all of the water has run through, repeat the process, but this time give the bottles a swirl.
- You should see an impressive spinning vortex or whirlpool effect, and the water will move rapidly from the top to the bottom bottle.

WHY does it work?

It's all about spin, and the same principle helps hurricane-force winds to accelerate up to speed.

When the bottles are first inverted, for water to move into the lower bottle, the space vacated in the upper bottle needs to be filled by air. Otherwise it would leave behind a vacuum. The bubbles of air rising through the narrow neck where the bottles are joined slow down the flow of the water so it takes a long time for one bottle to empty into the other.

However, when the bottles are swirled, the process is extremely rapid. This is because the water begins to spin, and as it moves downward toward the narrow neck, it's forced to turn in a tighter and tighter circle.

In the same way that pirouetting ice-skaters rotate much faster when they pull their arms in toward their bodies, confining the water to the narrow neck of the bottle accelerates the rate at which it is turning.

Quite quickly the water begins to spin so fast that it's flung outward against the sides of the bottle, creating a whirlpool effect with a hollow center. Air displaced from below can then easily move upward through this tube, helping to speed up the exit of the water.

HOW does this apply to the real world?

Hurricanes are triggered in the tropics by sea temperatures hotter than 83°F (28.5°C), that heat up air above the water, making it lighter (less dense) so that it begins to rise. At the same time, water vapor also evaporates from the sea surface, and this too begins to rise. As the warm wet air climbs, it leaves behind an area of lower pressure. This causes colder air to move in to fill the void, and the process is repeated.

Since the Earth is rotating, the cooler air that moves in is also turning. And just as the slowly spinning water speeded up as it was funneled into the narrow neck of the bottle, the same happens to the air in a hurricane, and it begins to turn much more quickly.

If the conditions are just right, the process builds up, gathering strength, until a hurricane is produced. In a large storm, like Hurricane Katrina, the wind speeds can exceed 140 miles per hour and tear off roofs, uproot trees, and cause massive sea swells.

The heart of the hurricane, which is known as the "eye," is like the tube of air in the middle of the vortex you created. Here, the wind, like the water in the bottleneck, is spinning so fast that it is flung outward away from the eye. This creates a very low pressure area at the center of the hurricane that draws in air from above. This descending air is dry and not turning, so there is very little wind, no clouds, and the sun is usually shining.

SOME OTHER THINGS TO TRY

You can make the effect even more dramatic by adding some colored oil (such as lamp oil from a craft shop) to the water in the bottles. The oil will float on the water, but when the vortex forms, it will flow down the center with the water around the outside.

IS THAT EGG HARD-BOILED?

How can you find out in 5 seconds if an egg is raw or hard-boiled without cracking the shell? This experiment shows you how, explains why this is important in industry and may even help you to achieve a hole in one on the golf course...

- ◆ You need a raw egg and a hard-boiled egg and a flat surface.

- ◆ Take the raw egg, lay it flat on its side and spin it.

- ◆ While it's spinning, lightly place a finger on top of the egg to stop the movement, and then quickly remove your finger.

- ◆ Surprisingly, the stationary egg will start to spin again, all by itself. You can even stop it several times and it will begin to turn again. Now try it with the hard-boiled egg. When this stops, it stays still. You should also notice that it's harder to start the raw egg spinning in the first instance, compared to the hard-boiled one.

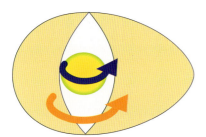

Hard-boiled egg, spinning

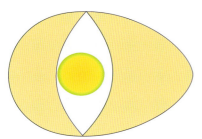

Hard-boiled egg, stopped

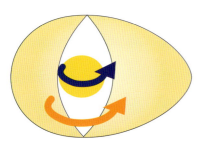

Raw egg, spinning

WHY does it work?

And what's the industrial connection? This effect is all down to the science of momentum, and it makes a big difference to how we move things around.

When an egg spins, both the shell and the contents turn. In the hard-boiled egg, because the contents are solid and fixed to the shell, everything turns together. This makes it easier to start the egg spinning to begin with and also makes it turn more stably, because everything moves around a fixed center of mass. When you stop the shell, you also stop the center, so the egg doesn't start moving again when it's released.

But in the raw egg, the contents are liquid. When the shell is stopped momentarily with a finger, although the egg appears to be standing still, the liquid continues to move inside, retaining its momentum. When it is released again, friction between the moving contents and the shell transfers some of the momentum back to the shell and it starts to turn once again.

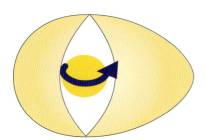

Raw egg, stopped

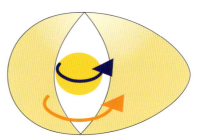

Raw egg, released

HOW does this apply to the real world?

This experiment is a good example of how fluids—being mobile—can build up momentum that can be a serious problem when it comes to transporting liquids aboard boats, trains, planes, or trucks. The movement of the vehicles is transferred to the liquids they carry, just as the shell transferred its rotation to the egg's contents in this experiment.

The movement of the liquid can cause the vehicle to become unstable. Imagine a large oil tanker rising and falling in a heavy sea. The oil in its tanks will try to follow the motion of the ship; this could lead to the cargo making exaggerated movements that apply stress to the hull and may even capsize the vessel. Or if a truck stopped quickly, the liquid it was carrying could cannon forward, potentially damaging the structure of the vehicle and making stopping more difficult.

To get around this problem, the tanks are divided into separate shorter sections to minimize the amount of movement the liquids can make. This also ensures that the weight of the cargo is evenly distributed along the length of the vehicle or vessel.

What's this got to do with scoring a hole in one? Some manufacturers produce golf balls with liquid cores to help the unskilled golfer stay on the fairway. Hitting a ball off-center causes it to spin and fly off course. But in the same way that the raw egg was difficult to spin because the yolk applied drag on the shell, a liquid-cored ball will spin much less, keeping it on the straight and narrow.

SOME OTHER THINGS TO TRY

Try spinning your hard-boiled egg really fast. If you can make it turn quickly enough, it will appear to defy gravity and stand up on its end. The reason for this was worked out recently by mathematician Keith Moffatt from the UK's Cambridge University. He found that because the side of an egg is a curved surface and the mass inside is not distributed evenly, when the egg rotates it tries to slip across the table. The most stable way for it to balance out the forces acting upon it is for it to stand up on its end.

MEASURE THE SPEED OF LIGHT WITH MARGARINE

While it might not be possible for a person to travel at the speed of light, you don't have to be a rocket scientist to measure how fast it goes. All you need is a ruler, some bread and margarine (or butter), and a microwave...

◆ Remove the turntable from your microwave.

◆ Arrange four pieces of bread in a square shape on a plate.

◆ Completely cover them with a thick layer of margarine, taking care to include the joins where the slices meet.

◆ Place the plate in the microwave. You need to make sure that your plate of bread won't turn when you switch on the microwave. This means that you might need to use an upside-down bowl to cover the central pillar that supports the turntable. If this is the case, balance your plate of bread on top of the upside-down bowl.

◆ Now cook the bread, on full power, for 15–20 seconds. Some microwaves are more powerful than others, so check every 5 seconds to see whether the margarine has begun to melt. When it does, you should see a series of parallel melted patches or lines in the margarine, separated by unmelted patches.

- Measure the distance in inches (centimeters) between two of these melted patches with the ruler, multiply this by two and write down your measurement. This is the wavelength of the microwaves produced by your oven. It should be about 4¾ inches (12 centimeters).

- Next, you need to find out the frequency of your microwave. This is the number of waves it produces per second. There should be a sticker, usually at the back or occasionally on the door lip, that gives the frequency either in gigahertz or megahertz. As a guide, or if you cannot find the value for your oven, most microwaves work at about 2,450 megahertz (2.45 gigahertz).

- Multiply the wavelength you calculated above by the frequency of your microwave. Take care with the units. If you are using megahertz, you need to multiply the answer by 1 million. If you are using gigahertz, you need to multiply your answer by 1 billion.

- In US measurements, the result will be the speed of light in inches per second. Convert it to miles per second by dividing the result by 63,360 (the number of inches in a mile). Then convert to miles per hour by multiplying by 3,600 (the number of seconds per hour). Your answer should be about 670 million miles per hour.

- In metric measurements, the result will be the speed of light in centimeters per second. Convert it to meters per second by dividing the result by 100. Your answer should be about 300 million meters per second.

WHY does it work?

Light, which includes microwaves, is a wave consisting of a series of peaks and valleys. One complete wave, known as the wavelength, is the distance from one peak to the next, while the frequency is how many waves are produced like this every second. To calculate how fast the wave is traveling, we need to know both of these values.

In a microwave oven, a chain of waves is produced on one side of the unit, reflects off the opposite side and then returns to where it started, overlapping with the original wave. In some places, the two waves cancel each other out, creating cold spots, but in other areas they add together, forming a hot spot.

The space between these hot spots corresponds to half the wavelength; in other words, the distance from a wave peak to the next valley. The hot spots also coincide with the places on the surface of the bread where the margarine melts first. Since the distance between a peak and a valley is half the wavelength, if you double the distance between two melted patches of margarine on your bread, this must be the wavelength. This is why you multiplied the distance you measured above by two.

Finally, the speed of the wave is calculated by multiplying the wavelength (the distance traveled) by the frequency (the number of times the distance was traveled in a second). Hey presto! You've measured the speed of light in your kitchen!

You've also shown very effectively why microwaves need a turntable, because the pattern of hot and cold spots wouldn't cook things evenly. By rotating food on a turntable, every part of it is exposed to a hot spot. This ensures that it heats up thoroughly.

Microwaves are produced by the magnetron on the right side of the oven (red) and reflect off the far side (black).

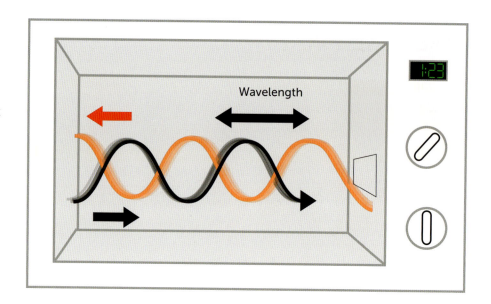

The outgoing and reflected waves overlap each other producing hot spots that melt the margarine.

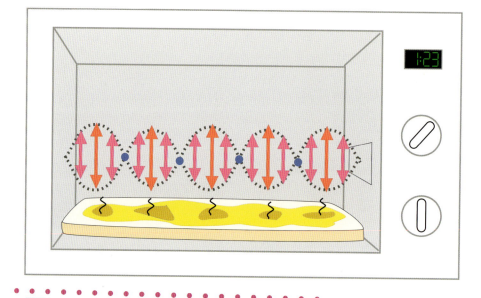

SOME OTHER THINGS TO TRY

A tastier alternative to bread and margarine is to try the experiment with squares of chocolate, or even marshmallows, both of which also give good results.

FREEZE A SOFT DRINK, INSTANTLY

Waiters often say "Ice with that?" But could you produce instant ice in a drink, merely by opening it? Yes, and in this experiment we'll show you how.

◆ To freeze a soft drink in front of your eyes, you'll need several small, previously unopened plastic soft drink bottles, an ice bucket or large bowl, lots of ice cubes, some salt, some water and, if you have one, a thermometer.

◆ Use your ice bucket to make an instant table-top freezer in the form of an "ice and salt sandwich." This consists of alternating layers of crushed ice cubes and salt. Make it sufficiently deep that you can bury the soft drink bottle in the ice.

◆ Add water to a depth halfway up the ice to improve the rate of cooling.

◆ With your thermometer, measure the temperature of the ice. The salt forces the ice to melt, driving down the temperature, and it should be possible to achieve at least 0°F (-18°C) using this method. In fact, before freezers were invented, people made ice cream this way, by cooling the cream in a bowl surrounded by an ice and salt mixture.

◆ Cool your soft drink to between 23 and 26°F (-3 and -5°C). You may have to do this by trial and error. As a guide, if you see ice beginning to form in the bottle, then it's become too cold!

◆ When the temperature is right, quickly open the bottle and allow the fizz to escape.

◆ Now watch. If the temperature is correct, within 10 seconds, the liquid in the bottle will turn to ice in front of your eyes.

WHY does it work?

It's down to the same science behind why we add antifreeze to engines, salt to the roads in winter and the phenomenon of freezing rain. The gas used to make soft drinks is carbon dioxide (CO_2). It dissolves in water to make a weak acid (carbonic acid) that has a slight lemon taste. This adds to the flavor and texture of the drink, which is why carbon dioxide is a popular additive.

Apart from its effect on the taste, adding carbon dioxide to water also affects its melting and boiling points. In fact, this is true when any chemical is dissolved in a liquid: the presence of the impurity makes the liquid turn into a solid (i.e. freeze) at a lower temperature and turn into a vapor (i.e. boil) at a higher temperature, compared with the pure form.

So the dissolved carbon dioxide reduces the freezing point of the water in the soft drink. This means that when carbon dioxide is present, the drink will remain as a liquid at a temperature lower than would normally be required to make it freeze.

However, when you remove the top, the carbon dioxide begins to escape, and this does two things. First, as the carbon dioxide bubbles out, it's no longer dissolved, so the freezing point of the liquid begins to rise. Second, the rising bubbles and froth help small ice crystals to form, and once small crystals are present it is much easier for other crystals to begin to grow. This is known as nucleation. The result is that, once open, the drink freezes very quickly from the top down, where most of the nucleation took place in the froth.

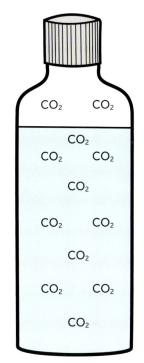

Closed: freezing point reduced by dissolved CO_2.

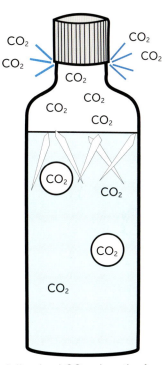

Open: loss of dissolved CO_2 raises the freezing point, allowing the drink to form ice crystals.

HOW does this apply to the real world?

During cold weather, trucks sprinkle salt onto roads to stop them from freezing. The principle is identical to the soft drink in this experiment. The salt dissolves in the water on the road surface and lowers its freezing point. The more salt that dissolves, the colder it needs to be for the road to freeze. Similarly, when mechanics add antifreeze to the cooling water in an engine, they are dissolving a chemical in the water to lower its freezing point and to reduce the risk of it turning to ice, which could rupture the radiator and hoses.

Nucleation also produces the freaky weather phenomenon of freezing rain—liquid rain that instantly freezes the second it hits the windshield of a car. This can be extremely dangerous because it can very quickly make it impossible to see the road ahead. This effect happens when raindrops become super-cooled below 32°F (0°C) by a region of very cold air as they fall. They don't freeze in the air because the absence of any irregular surfaces makes it very difficult for ice crystals to begin to form. Then, when they land on the car windshield, imperfections on the surface act as nucleation sites—like the froth and bubbles in the soft drink in this experiment—and the drops freeze.

SOME OTHER THINGS TO TRY

Try pouring the drink out into a glass, or hitting the side of the bottle hard to nucleate the ice and trigger freezing.

CHIP BAG/CRISP PACKET FIREWORKS

Sometimes you see a stunning display of sparks and fireworks when a prized piece of gold-edged porcelain ends up in the microwave. In this experiment, we'll recreate the effect much more cheaply and also find out why you shouldn't (in theory) use a cell phone at a gas station.

ADULT SUPERVISION!

◆ For this experiment, you'll need a microwave and a foil chip bag/crisp packet.

◆ Place the bag in the microwave on an old plate.

◆ Cook the bag for, at the most, 4–5 seconds and watch what happens.

◆ Turn off the microwave and then wait a minute or so for things to cool down before you remove the bag.

◆ If the experiment worked correctly, you will have seen the bag shrinking rapidly, producing an impressive display of sparks. Afterwards, the plastic should have become much thicker and stiffer.

WHY does it work?

This experiment is all about the behavior of polymers and the way in which microwaves can induce an electric current in metal objects.

"Foil" chip bags/crisp packets like the one used here consist of a thin layer of aluminum foil sandwiched between two thin layers of plastic. The aluminum is there to prevent oxygen and sunlight from spoiling the contents.

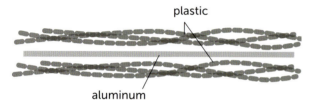

plastic

aluminum

Aluminum conducts electricity, so when it's hit by microwaves (which are a form of electromagnetic radiation), a current begins to flow back and forward in the metal. This causes the metal to heat up, and some of the heat passes into the plastic on either side.

When the microwaves hit the aluminum, they induce electrical charges to flow backward and forward in the metal, producing a current.

Plastic is a polymer made of long chains of small molecules linked end to end, rather like a string of beads. To make a chip bag/crisp packet, the molecules are stretched out into a thin film. When they become hot, due to the effect of the microwaves on the metal, the molecules begin to vibrate. This allows the stretched out polymer chains to slide past each other and form short fat tangles. This is why the chip bag/crisp packet shrinks and becomes thicker.

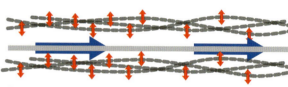

The current (blue) travels through the aluminum, heating the plastic and causing the polymer molecules to vibrate (red).

What about the sparks? This is due to the electrical current induced in the aluminum foil by the microwaves. As the plastic shrinks, it causes cracks and crinkles to form in the foil. This causes the flowing current to stop suddenly, and a large amount of electrical charge piles up behind the crack.

If enough charge accumulates, it can overcome the resistance of the air and jump from one side of the crack or crinkle to the other as a spark. When this happens, the surrounding air becomes very hot, causing it to glow and expand rapidly. This produces a miniature shock wave that you hear as a crackling sound.

The polymers form short, fat tangles, introducing cracks in the foil. This interrupts the current and it jumps the gap as a spark.

HOW does this apply to the real world?

This experiment highlights two important principles. One is how "thermoplastic" polymers work. When they are heated to a sufficiently high temperature, the polymer chains can flow past each other so that the material can be molded into a desired shape. When they cool down, the chains can no longer flow past each other, so the plastic retains its new form.

The second principle is how electromagnetic waves, such as microwaves, can induce an electrical current in metal conductors. This is how radio waves, television, and telephone signals are transmitted and received. When an arriving radio wave hits your antenna, it induces an electrical current that the radio can amplify and turn into sound waves that you can hear.

It's on this basis that the owners of gas stations ask you not to use a cell phone at the pump due to the theoretical risk that the microwaves emitted by the phone might induce a current in a nearby metal object, triggering a spark that could ignite gasoline vapor. That said, the authors have seen more sparks from someone removing their sweater than using their phone, but we don't make the rules!

SOME OTHER THINGS TO TRY
Thin foils of aluminum are also used in CDs, so you could try the experiment with a disc you don't like.

HOMEMADE FIBER OPTICS

Whenever you surf the Internet or make a telephone call, it's almost certain that the information you are sending and receiving is traveling, for at least part of its journey, as light pulses in fiber-optic cables. But how do fiber optics work and why are they better than pieces of wire?

◆ For this experiment you will need an empty plastic drink bottle, a bright flashlight, a drill or something similar with which to make a hole in the bottle, some water, and a sink or bathtub.

◆ Start by making a small quarter-inch (half-centimeter) hole in the side of the bottle close to the bottom.

◆ Put your finger over the hole to block it and fill the bottle with water.

◆ Then, position the flashlight so that it shines through the bottle from the side opposite the hole.

◆ Now, place the bottle over the sink or bathtub, and remove your finger from the hole so that a jet of water emerges and curves downward into the sink.

◆ Place your hand in the stream of water at various places along its course below the level of the bottle. What you should see is a spot of light on your hand where the water hits you.

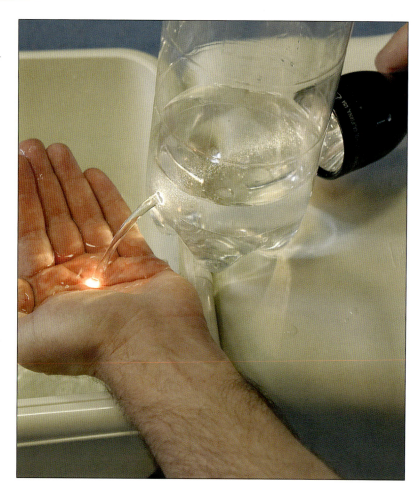

WHY does it work?

Why is the light trapped inside the curved jet of water? This is the principle of total internal reflection, that is vital for fiber optics. It also involves the same science that makes a drinking straw look bent when you dip it into water.

To understand what's going on try this simple experiment. Take a glass bowl, fill it with water and lay a spoon on the bottom. Now look from below through the side of the bowl at the surface of the water. You should be able to see the spoon reflected in the underside of the water surface. In other words, the water is behaving like a mirror.

When the stream of water leaves the hole in the side of the plastic bottle, light entering

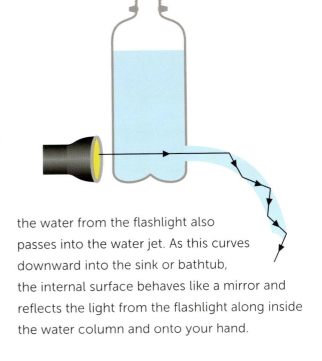

the water from the flashlight also passes into the water jet. As this curves downward into the sink or bathtub, the internal surface behaves like a mirror and reflects the light from the flashlight along inside the water column and onto your hand.

HOW does this apply to the real world?

Light changes speed when it passes from one substance into another, such as from water into air, and this causes the path of the light to bend, which is known as refraction. This is why a drinking straw sometimes appears to be oddly kinked at the point where it enters the liquid in a glass of water.

When light moves from water into air, it bends toward the surface of the water, and the shallower the angle at which it hits the surface the more it bends. So at a shallow enough angle, the beam of light should be bent so much that it would stay inside the water. However, to have refracted, the light must have traveled into the air, when of course it hasn't. So the light solves this paradox by instead

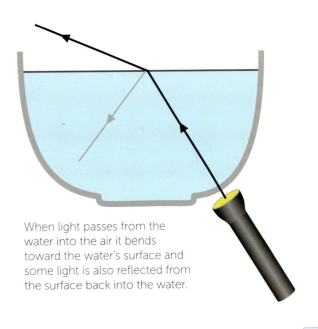

When light passes from the water into the air it bends toward the water's surface and some light is also reflected from the surface back into the water.

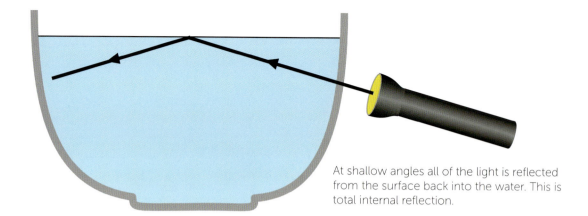

At shallow angles all of the light is reflected from the surface back into the water. This is total internal reflection.

reflecting off the underside of the water's surface, that behaves like a perfect mirror. This is why, in this experiment, you could see the spoon reflected in the surface of the liquid when you looked through the side of the bowl from below.

This phenomenon is called total internal reflection, and optical fibers work the same way. They consist of a fine thread of very high-purity glass surrounded by a slightly different glass. Light travels slightly faster in the outer glass layer, so it acts like the air around the water jet in this experiment.

As light is channeled down the optical fiber, it meets the junction between the two glass layers at a sufficiently shallow angle that it is totally internally reflected off the sides, even as it bends and twists around corners.

To send information along an optical fiber, a computer first converts electrical data into very short pulses of light that are injected into the fiber by a light-emitting diode (LED) or a laser. They travel along the fiber and are then "read" at the other end by a light detector, that converts

the light back into electrical signals that the computer can understand.

You can transfer a huge amount of information this way. The present record is over 250 terabits every second, and using very high-quality-glass optical fiber you can send the information tens of miles before you have to amplify it again. As a result, the bulk of the world's communications, including most of the Internet, now travel this way.

A more festive application of this technology is in artificial self-illuminating Christmas trees that don't need holiday/fairy lights. Instead the "tree" contains a mass of plastic optical fibers linked to a light source in the base. These plastic fibers funnel the light up the tree, where they then illuminate the branches.

SOME OTHER THINGS TO TRY

If you go swimming when the water is quite calm, gently dive down and look up at the surface. At an angle of less than 40 degrees from the horizontal, the surface should look like a mirror because of total internal reflection.

SOFT DRINK VOLCANO

How do volcanoes work, and why do some erupt explosively, blowing themselves to pieces? In this experiment we'll make a volcano that can erupt safely outdoors and will explain why gas is important to the process.

STAND WELL BACK!

◆ You will need several unopened soft drink bottles, a supply of mints (such as a packet of Mentos™ or strong mints), and a piece of paper that can be rolled into a tube.

◆ Place a soft drink bottle in a clear area outside and gently open it, minimizing the amount of gas and liquid that escapes.

◆ Now, roll up the piece of paper to make a tube the correct size for your mints. Alternatively, if your mints have a hole through the middle, you can slide them onto a piece of wire. Whichever method you use, you need to be able to add all of the mints at once to the open bottle, and extremely rapidly. The more mints you can add, the more impressive the result.

◆ As soon as they enter the bottle, stand back. The liquid will explode through the narrow neck of the bottle and shoot up to three feet (a meter) into the air.

WHY does it work?

It's all down to dissolved gas and a process known as nucleation; it's also the driving force behind some of the world's worst volcanic explosions.

To produce soft drinks, manufacturers force carbon dioxide gas into the liquid under pressure, and the gas dissolves. With the top on, the pressure inside the bottle prevents the gas from turning back into bubbles, so the liquid remains fizzy.

When you open the bottle (without adding any mints), and the pressure drops, bubbles find it easier to form, and some will trickle to the surface. But water is a sticky molecule, and to form a bubble inside the liquid, the gas has to overcome this stickiness and push water molecules apart. Small bubbles therefore find it difficult to form in the first place.

Nevertheless, trapped within the rough surfaces of the mints are millions of tiny pockets of air. When the mints are submerged in the soft drink, these air pockets become tiny bubbles that then "seed" the formation of larger bubbles filled with carbon dioxide taken from the drink.

And because the mints sink to the bottom of the bottle, large numbers of bubbles, that take up large amounts of space, form underneath the liquid. This pushes everything upward and out through the neck of the bottle. Since the neck is narrower than the body of the bottle, the liquid has to speed up as it passes through, which is what sends the drink shooting up into the air.

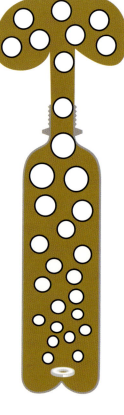

HOW does this apply to the real world?

Volcanoes are powered by accumulations of molten rock, known as magma, that build up beneath the Earth's surface. The magma is under enormous pressure that forces any gases that are present, including carbon dioxide, sulfur dioxide, and water vapor, to dissolve in the liquid rock. If the pressure suddenly drops, for instance at the onset of an eruption, the dissolved gases form bubbles thousands of times larger than the original volume of magma. This blasts molten rock, pumice, dust, and ash for miles in all directions and often blows the volcano itself to pieces. A dramatic example of this happening in recent times is the 1980 eruption of Mount Saint Helens in Washington, USA.

SOME OTHER THINGS TO TRY

You can make the effect more dramatic by making the neck of the bottle even narrower. An easy way to do this is to make a small hole in the top. Screw this back on quickly after adding the mints. (You are advised to do this outside because the jet it produces can be several feet [meters] in height).

TOASTER-POWERED HOT-AIR BALLOON

On a warm evening, you often see hot-air balloons floating along. But what keeps them aloft? In this experiment, you'll see that ballooning should be the preferred pursuit of politicians...because it requires a lot of hot air.

ADULT SUPERVISION!

- You will need a kitchen toaster, a thin polythene plastic bag of the sort that is used to line a trash bin, some adhesive tape, and a length of cardboard similar in height to the plastic bag and large enough to encircle the toaster.

- Roll the cardboard into a rough tube large enough to fit over the toaster but small enough to fit inside the bag.

- Fix the edges of the cardboard together using the tape so that it looks like a large chef's hat. (Make sure that you apply the tape to the outside of the cardboard rather than to the inside where it could touch the toaster.)

- Switch on the toaster and place the cardboard tube over the top so that it resembles a chimney.

- Now, open the bag and place it over the cardboard tube and then wait.

- After 15–20 seconds, the bag will take off. Once the bag has cleared the launch pad, switch off the toaster. With a clear flight path, it should easily reach a height of 6–10 feet (2–3 meters) before it begins to drift down again.

WHY does it work?

Hot air balloons float for precisely the same reasons as boats: it's all down to density. Hard as it is to believe, the air around us is heavy. At sea level, each cubic foot (cubic meter) weighs about 1 ounce (1 kilogram), which means that the air in your trash-bag balloon (prior to take off) weighs about 2½ ounces (70 grams).

When the toaster is turned on, it heats the air inside the bag, causing it to expand. This means that there will no longer be enough space in the bag for all of the original air to fit. As a guide, if the temperature rises by 54°F (30°C), it will cause the air to expand by about 10 percent. So, roughly ¼ ounce (7 grams) of air (10 percent of 2½ ounces [70 grams]) will be pushed out of the bag when the toaster is switched on.

Since the plastic used to make the bag weighs only about ⅕ ounce (5 grams), and it has lost air weighing ¼ ounce (7 grams), the balloon-bag now weighs less than the surrounding air, which tries to sink below the balloon, pushing it upward in the process.

Another way to look at it is that since a balloon full of hot air is the same size as a balloon full of cold air but weighs less, it must be less dense. Things float when they are less dense than the air (or water) around them. The balloon will continue to rise until it reaches air of the same density as itself.

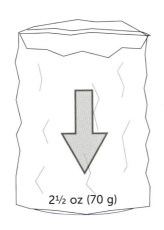

2½ oz (70 g)

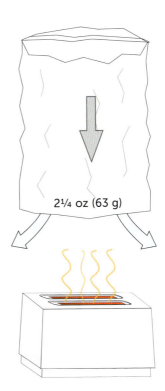

2¼ oz (63 g)

HOW does this apply to the real world?

The principle behind this experiment is identical to what happens aboard the hot-air balloons you see drifting along in the sky. The only difference is that modern balloons don't use a toaster. Instead, they burn propane gas stored in cylinders to heat the air inside.

The first hot-air balloons were invented in the 1700s by the French Montgolfier brothers, who used a fire beneath the mouth of the balloon as the heat source. Their first passengers, in 1783, were a sheep called Montauciel ("Climb to the Sky"), a duck, and a rooster, and subsequently a physician called Pilâtre de Rozier and an army officer, the Marquis d'Arlandes. They achieved an altitude of about 3,000 feet (915 meters), but now the record for the highest balloon ascent stands at over 60,000 feet (18,000 meters). It was set by an Indian textile tycoon, Vijaypat Singhania in November 2005.

SOME OTHER THINGS TO TRY

You can try adding some cargo to your balloon to see how much weight it can carry upward. Use adhesive tape to attach small balanced payloads to the bottom of the bag.

You can also try holding the bag down before you let it fly. This will make the air inside hotter, and therefore lighter so the bag will lift much faster—but be careful not to melt the plastic!

USE YOUR LOAF–HOW TO MAKE BREAD TASTE SWEET

The human body is effectively a mobile bag of chemical reactions. Most of these reactions are controlled by biological catalysts called enzymes, but what do they do and how can they help to remove stubborn stains from your washing?

◆ For this experiment, you'll need a slice of white bread (the cheaper the better), and a willing mouth to put it in.

◆ Chew the bread, but do not swallow it. Just keep chewing, paying careful attention to the flavor of the bread as you chew.

◆ It may take 5 minutes before you notice anything, but eventually the bread will begin to taste sweet, like sugar.

WHY does it work?

The sugary sensation is down to the action of enzymes in your saliva converting one molecule into another, right in front of your taste buds. Bread is made from flour, which is mostly starch. This is the plant equivalent of human body fat; it's how plants store the energy they produce when they use their green pigment chlorophyll to photosynthesize in sunlight. The main products of photosynthesis are glucose, which is a sugar, and oxygen, which we breathe.

Glucose dissolves in water. This makes it easy for plants to transport it from the leaves, where it's made, to other parts of the plant where it can be stored, such as in tubers (like potatoes) or in seeds (like ears of wheat).

Once it arrives at its destination, the plant needs to keep it there in a space-efficient compact form that won't soak up lots of water. To do this, the glucose is converted into starch. This is a plant polymer made by joining thousands of individual glucose molecules together to form long chains. For added resilience, the chains are then cross-linked together.

Not surprisingly, these massive molecules won't dissolve in water, so they're easy to store. When the plant or a germinating seed needs energy, the starch can be broken up again to release the individual sugar molecules.

This is exactly what is happening in your mouth when the bread begins to taste sweet. Saliva

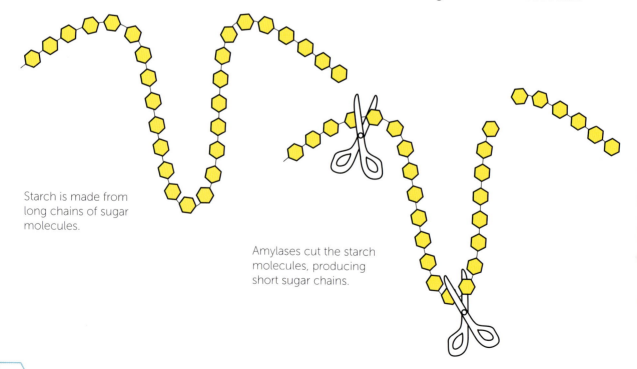

Starch is made from long chains of sugar molecules.

Amylases cut the starch molecules, producing short sugar chains.

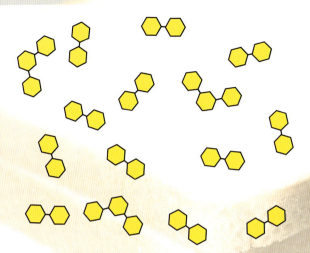

Ultimately, the sugar chains are cut into two- and three-sugar units called maltose and malto-triose, that taste sweet.

contains small amounts of an enzyme called amylase (or ptyalin) that cuts up starch (another name for starch is amylose) into smaller sugar molecules called maltose and malto-triose. These dissolve in the saliva and lock onto sugar-detecting taste buds on the tongue, making the bread taste sweet.

Why do we have this enzyme in our mouths? No one knows for sure, but some scientists think that it might work a bit like a chemical toothpick, dissolving pieces of starch stuck between the teeth.

HOW does this apply to the real world?

Enzymes are chemical catalysts made in the cells of animals, plants, bacteria, and fungi using a recipe written into the cell's DNA. They enable chemical reactions to take place at much lower temperatures and much more quickly than normal. The body uses enzymes like amylase to break down the foods we eat so that we can digest nutrients. Giant molecules, such as starch and most proteins, are too large to be absorbed and cannot be used in the body. Instead they're chemically dismantled into their component parts, again using enzymes that are released into the digestive juices.

The process is rather like sending a car to the junk yard for recycling. Once it gets there, all of the useful parts are stripped off and taken away for use in other cars. What's left gets thrown away.

The smaller molecules produced by the digestive process are picked up by cells along the intestine and added to the bloodstream. Other cells around the body then retrieve the components they need and use yet more enzymes to construct new molecules from these chemical building blocks.

Enzymes are also used to clean laundry. Many hard-to-remove stains like egg, blood, and tomato contain long protein molecules that anchor themselves tightly to clothing. Normal laundry detergents struggle to free them at low temperatures, but biological laundry powders contain enzymes that can break down the proteins, so the stain can be washed out, even at low temperatures.

WHY IS THE SKY BLUE?

At night, the moon and stars are white, meaning that the atmosphere must be colorless. So why does the sky look blue during the day, and why does the Sun turn red at sunset?

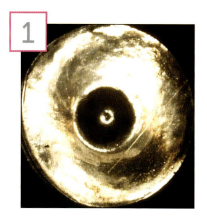
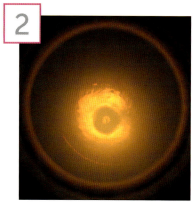
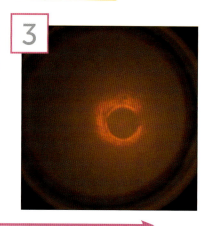

Increasing depth of liquid

◆ For this experiment, you will need a large glass or jar, a jug, a flashlight, some milk or milk powder, and some water. It also works best in a dark room.

◆ Start by filling the glass with water and then empty it into the jug.

◆ Mix a very small amount of milk or milk powder (1–2 level teaspoons should be sufficient) into the water and pour some of the mixture (to a depth of 2–3 inches [several centimeters]) back into the glass.

◆ Shine the flashlight through the base of the glass, and look through the liquid from above.

◆ What color is the bulb? Looking from the side of the glass, what color is the liquid?

◆ Now add more liquid from the jug and repeat the process.

◆ Keep going like this until the glass is full or until you can no longer see the bulb through the liquid. If the bulb disappears almost immediately, you have probably added too much milk, and you should dilute the solution with more water.

◆ If it's working correctly, as the glass fills up you should see the filament of the bulb change color. Starting off white, it should become more yellow, then orange, and eventually red. Looking from the side, the liquid should have an orange hue at the top, a yellow hue midway up, and the base should look slightly bluish (you may have to dilute the mixture to get this to work).

WHY does it work?

It's down to the "Tyndall Effect," a phenomenon first described 150 years ago by the Irish scientist John Tyndall, who discovered how particles scatter light.

Although sunlight and the light from flashlights, headlights, and fluorescent tubes look "white," Isaac Newton showed in the 1600s that white light is actually made up of a whole spectrum of colors. You can see them whenever there's a rainbow, or if you split up white light with a prism.

These colors are electromagnetic waves, like radio waves, and different colored lights have different wavelengths. The bluer colors have shorter wavelengths, while yellow and red colors have longer wavelengths. Adding them together produces the ultimate color of light that we see.

What Tyndall showed is that when light passes through a collection of particles, the shorter wavelengths are more easily deflected and scattered than the longer wavelengths.

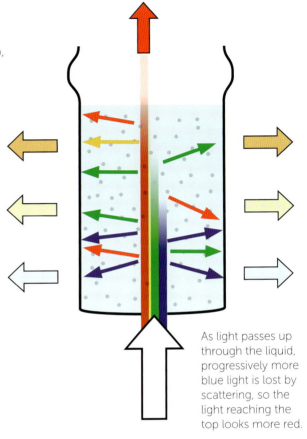

As light passes up through the liquid, progressively more blue light is lost by scattering, so the light reaching the top looks more red.

Milk particles can work the same way and scatter the shorter (bluer) wavelengths from the flashlight more than the longer (redder) wavelengths. As the light passes through the liquid, progressively more of the blue and green colors are deflected out of the sides of glass, leaving only the red and yellow light to make it through to the top. From the side, this makes the bottom of the glass look bluish, and the top look redder.

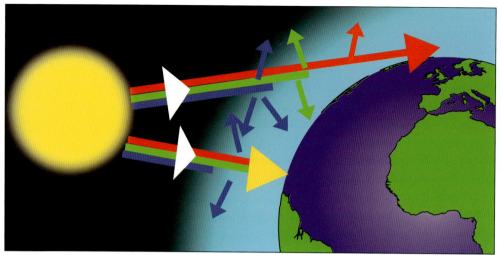

As light travels a greater distance through the atmosphere at sunrise and sunset, more blue light is scattered, making the Sun appear red.

HOW does this apply to the real world?

The same phenomenon is at work in the sky, although here the effect is mostly caused by molecules of oxygen and nitrogen in the Earth's atmosphere, rather than by particles of milk.

When white light arrives from the sun, some of the short-wavelength blue light is scattered and bounced about in all directions so that it no longer appears to be coming from just one place, the Sun. Instead, the whole sky seems to be blue. The rest of the light with longer wavelengths passes on unaffected, which is why the Sun looks a yellow color—white minus some of the blue.

Why does the Sun look red on the horizon? As the Sun sinks, or when it first rises, the light has to travel a greater distance through the Earth's atmosphere. As a result, more of the shorter blue and green wavelengths are scattered, making the Sun look redder.

The effect is even more noticeable when there is a large amount of dust or pollution in the atmosphere, such as after a volcanic eruption or during the summer crop harvest. Airborne dust behaves just like the milk particles, making the sun look redder on the horizon.

This is why, in this experiment, once the glass was full, the light emerging from the top was a red color: the majority of the blue and green light had been scattered.

You can also see the Tyndall Effect in action closer to the ground whenever you go driving on a foggy night. The beams of the headlamps will make the fog look blue. The lights of oncoming cars, however, look redder.

SOME OTHER THINGS TO TRY

If you can find some Polaroid sunglasses, have a look at a blue sky and your glass through the Polaroids. Now rotate them sideways. Do you notice anything happening?

EXTRACT DNA FROM A KIWIFRUIT

YUCK! DON'T EAT!

Whether you look at bacteria or buffaloes, kangaroos or kiwifruit, DNA is the cellular recipe book used by all living things on Earth. In the time it takes you to do this experiment, your body will have made miles of it. What does it look like? Here we'll show you a simple way to extract some of nature's blueprint from a common fruit.

1

- For this experiment, you will need a kiwifruit, some dishwashing liquid, some table salt, water, a bottle of strong alcohol such as rubbing alcohol/surgical spirit or aftershave, a coffee filter or a very fine sieve, a cereal bowl, a large heatproof glass bowl, a jam jar or wineglass, and a fork.

- Before you begin, put the rubbing alcohol/surgical spirit or aftershave in the freezer. It needs to be ice cold for this experiment to work.

- Start by peeling the kiwifruit; you can discard the skin.

- Cut the flesh of the fruit into small pieces and place them in the cereal bowl.

- Using the fork, thoroughly mash the pieces, and then add 1 tablespoon of dishwashing liquid, 1 teaspoon of table salt, and 3½ fluid ounces (100 milliliters) of water from the faucet/tap.

- Mix this in gently but firmly, continuing to mash the fruit for at least 5 minutes. Try to avoid producing excessive amounts of foam, as this will make the results harder to see.

- Next, keep the mixture at 140°F (60°C) for 15 minutes. This is best done in a water bath. You can make a simple one by half-filling a heatproof glass bowl with boiling water from the kettle and then adding an equal amount of cold water from the faucet/tap. Float the cereal bowl on the water in the large bowl.

- Stir the mixture gently from time to time.

- After 15 minutes, pass the contents of the cereal bowl through the coffee filter (or the fine sieve) to collect just the liquid into your wineglass. You need about half a glassful of liquid.

- Retrieve the ice-cold alcohol from the freezer and very gently pour the cold liquid down the inside edge of the glass.

- A layer of alcohol will form above the kiwifruit extract.

- Watch carefully where these two layers meet. Over a few minutes, a white stringy substance will begin to appear. This is kiwifruit DNA, and you should be able to hook it out with a dinner fork!

WHY does it work?

It's because every living cell in a plant contains at least one and sometimes several copies of the plant's DNA genome. Plant cells are surrounded by a rigid cell wall made from a polymer called cellulose. Inside the cell wall is an oily layer called the membrane, that surrounds the entire contents of the cell. The membrane is the cellular equivalent of Gore-Tex™; it restricts what substances can flow in and out in order to maintain the perfect chemical environment inside the cell. It also holds in the DNA; this is locked away in a structure called the nucleus.

Adding dishwashing liquid causes the cell membranes to dissolve, while heating the mixture to 140°F (60°C) breaks apart the cellulose in the cell walls. This allows the DNA, that isn't harmed by either of these processes, to escape into the solution, where the table salt encourages it to form clumps.

DNA won't dissolve in alcohol. Alcohol dehydrates the DNA chains, making them stick together. When you add the ice-cold rubbing alcohol, the DNA begins to precipitate out in the layer where the two solutions mix. The longer you leave it, the more DNA you see appearing.

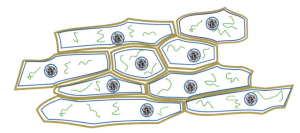

Kiwifruit cells are surrounded by a tough cell wall and an oily cell membrane.

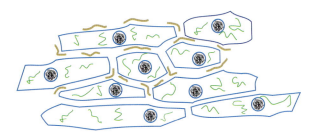

Heating the kiwifruit helps to break down the cell walls.

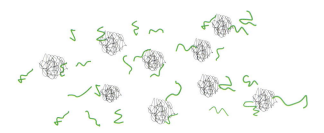

The detergent breaks down the cell membrane releasing the cell contents, including DNA.

Alcohol pulls water away from the DNA, causing strands to clump together.

HOW does this apply to the real world?

Scientists use a similar method to extract and purify DNA for analysis in the laboratory. This includes isolating human DNA, because many countries around the world are now setting up DNA databases to help the police to catch criminals. To do this, samples are collected by swabbing the inside of a person's mouth.

Cells brushed from the cheek and tongue are then treated with detergents to break them open and release their DNA; this can then be analyzed. This can include producing a genetic "fingerprint," that can be kept on record and then compared with samples collected at crime scenes.

SOME OTHER THINGS TO TRY

Try the experiment with other fruit and vegetables, or even extract some of your own DNA by gently chewing the inside of your cheek and spitting the cell-rich saliva into a pot. Then precipitate your genome by adding some detergent to break open the cells and some ice cold alcohol to make the DNA appear.

HOW TO FOOL YOUR SENSES

Have you noticed how when you first smell something the scent seems very strong but then it appears to weaken? Or when you first jump into the sea the water feels cold but then seems to warm up? Of course, neither is really happening, it's just your nervous system playing tricks on you. Why does it do this?

◆ For this experiment, you will need three bowls or buckets, some ice cubes, and some hot and cold water.

◆ Arrange the three bowls (or buckets) side by side and within arm's length of each other.

- To the left-hand bowl add the ice cubes and pour in cold water to a depth of 2 inches (5 centimeters). Be sure to allow enough time for the water to become really cold before you start the experiment.
- Then, to the middle bowl add some lukewarm water and to the right-hand bowl some hot (but not scalding) water.
- Once you are ready to start, place one hand in the cold water in the bowl on the left and the other hand in the hot water on the right.
- Leave them immersed for at least a minute and then quickly move both hands, at the same time, to the middle bowl of lukewarm water.
- Despite the fact that both of your hands are now touching water that is the same temperature, one hand should be telling you that the water feels hot but on the other hand (literally), the water feels cold! You may also notice that the hand that was previously in the cold water has now gone red.

WHY does it work?

This is down to adaptation, the way that the nervous system prevents sensory overload. The red hand, meanwhile, is due to a protective mechanism that keeps you warm on a cold day.

Temperature information is sent to the brain in the form of signals from heat-sensitive nerve endings in the skin. There are warm-detecting nerves that signal when the temperature is rising and cold-sensitive nerves that pick up when the temperature is dropping.

Both forms of nerve fiber contain heat-sensitive chemicals that alter their shapes with changes in temperature. This is used to control the activity of the nerve cell and the number of impulses that it sends to the brain.

These cells do not record absolute temperatures. Instead, they measure how the temperature is changing. The bigger the change, the more active they become. After an initial surge of activity they begin to switch off again, almost as though they have lost interest in the situation.

When you transferred your hand from the cold to the tepid water, the warm-sensing nerves in the skin became much more active, and any residual activity in the cold-sensing nerves was shut off. This fooled your brain into thinking that the water was hot.

For the hand that had been in the hot water previously, the situation was reversed. The sudden drop in temperature activated the cold-sensing nerves and shut off any residual activity in the warm-detecting cells. This led the brain to believe that the water must be cold. Thankfully, after a short while, both hands begin to feel the same as the nerve responses die down and agree with each other again; this is the process of adaptation.

And the red hand? This is due to a reflex that protects the body from excess heat loss. When a part of the body becomes very cold, blood vessels supplying the skin and surface tissues in that area constrict. This cuts down the flow of warm blood near the surface and reduces the rate of heat loss. You may have noticed your hand becoming quite pale as it sat in the cold water.

Tissues rely on the blood flow to deliver oxygen and energy and to remove waste products. If the blood supply is reduced for any reason, cells can run short of energy, and waste can build up. As a result, when the blood supply returns, more blood than normal is needed to replenish the cells with energy and to remove the excess waste.

When you put your cold hand in the tepid water, the constricted vessels opened up again, this time wider than normal, and blood surged in, making the skin look red. This is known as a reactive hyperemia.

HOW does this apply to the real world?

Without the process of adaptation we would be overloaded with sensory information and find it very hard to discriminate important signals in the environment. It allows us to get used to smells that have been around for a while (thankfully, in some cases, quite quickly) and explains why we are not continuously aware of the clothes we're wearing.

SOME OTHER THINGS TO TRY

Instead of placing your hands in the lukewarm water after the initial step in the experiment, just swap them over. Put the hot hand in the cold water and vice versa. Is the effect magnified? A similar sensory illusion can be created by gently rubbing one hand along some carpet and the other on something very smooth for a minute or so. Then rub both hands on some writing paper. It should feel rough and smooth at the same time!

ELECTRIC SLIME

Is it possible for a liquid to behave like a solid, just with the wave of a balloon? The answer is yes, and the same science can create a family of materials called ferrofluids that alter the way they behave whenever there's a magnet nearby.

- You will need some cornstarch/cornflour, some vegetable (cooking) oil, a plastic spoon, and a balloon.

- Begin by mixing adequate amounts of the cornstarch/cornflour and vegetable oil to produce a paste with the consistency of thick cream.

- Inflate the balloon and rub it on your hair or a sweater to charge it up with "static" electricity. When you do this, negative charges from your hair are added to the balloon, making it negatively charged as well.

- Now take a spoonful of the cornstarch/cornflour and oil mixture and bring the balloon close to it as you pour the mixture off the spoon.

- Watch the cornstarch/cornflour carefully. You should see the mixture move toward the balloon and become much thicker and stiffer.

WHY does it work?

This effect is due to electrostatics and the way unlike charges attract each other, and like charges repel. Cornstarch/cornflour consists of tiny particles of starch measuring less than one thousandth of an inch (one hundredth of a millimeter) across. Spread evenly throughout each of these particles are equal numbers of positive and negative charges. When the negatively charged balloon is brought near, the

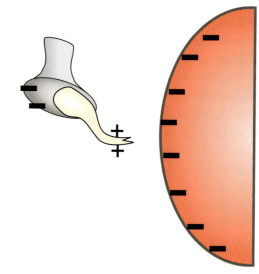

charges in the cornstarch/cornflour particles are forced to rearrange themselves.

The negative charges, that are strongly repelled by the negative charge on the balloon, move to the sides of the particles furthest from the balloon. This leaves the sides of the particles closest to the balloon with a net positive charge.

Since the positive charges in the particles are now closer to the balloon than the negative charges, there will be a net attractive force pulling the particles toward the balloon. This is why the cornstarch/cornflour moves toward the balloon.

Why did we mix the cornstarch/cornflour with oil? This is because oil is an insulator. It surrounds each of the particles and electrically isolates them from each other so that charges

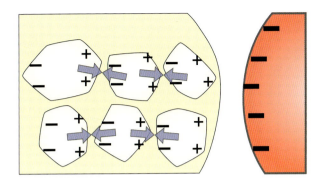

cannot move between them. As a result, when the balloon is brought near, the positive charges on one side of a particle strongly attract the negative charges of the particle next to it. This alters the consistency and increases the "stiffness" of the slime. Without the oil, or if water had been used instead, the charges would be able to move from one particle to the next so they would not attract each other in this way.

HOW does this apply to the real world?

The behavior of the electric slime is a good demonstration of how the consistency and behavior of a fluid can be altered using an electromagnetic field. Engineers can now make materials known as "ferrofluids," that are oily liquids containing millions of tiny suspended magnetic particles. When these fluids are exposed to a magnetic field, the particles line up with the field and pull the fluid into a new shape and texture. This technology has been used to seal the bearings of hard disk drives in computers to prevent the discs from becoming contaminated with particles from the bearings as they spin.

SOME OTHER THINGS TO TRY
Repeat the experiment, but this time use water instead of vegetable oil. Does it still work?

THE CHEMISTRY OF COPPERS

Why is it that newly minted coins are always bright and shiny but after a while they lose their luster and become dark and tarnished? This is quite literally dirty money, and it's the fault of oxygen.

- For this experiment, you will need a handful of well-tarnished copper coins and some white-wine vinegar or lemon juice.

- Pour a little vinegar or lemon juice into an egg- or teacup and add the copper coins. Try to prop one of the coins against the side of the container so that half of the coin remains above the level of the liquid.

- Wait 10 minutes and then remove the coins and dry them on a piece of kitchen papertowel.

- What's changed? You can use the coin that was only half-immersed for comparison. They should have gained a shiny new lease of life and have shed all of their tarnish. Now there's money laundering for you!

WHY does it work?

This is the chemical consequence of exposure to acids that can be used to clean up surfaces and strip away unwanted oxides. The technique can even be used to cure a rusty car. The tarnish that forms on copper coins is caused by the metal reacting with oxygen in the air to form copper oxide. This is a black color, which is why coins darken as they age—because the thickness of the oxide layer increases. The oxide can be removed by adding an acid, such as vinegar, that contains acetic acid, or

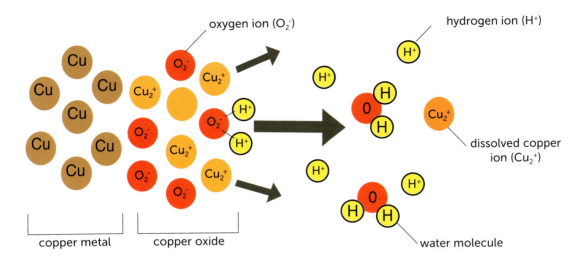

oxygen ion (O_2^-)

hydrogen ion (H^+)

Cu

Cu   Cu Cu_2^+ O_2^- Cu_2^+

Cu Cu_2^+

Cu O_2^- O_2^- H^+

Cu   Cu Cu_2^+ H^+

Cu O_2^- Cu_2^+

O_2^-

H^+

H^+

H^+ H
       O
          H

Cu_2^+

dissolved copper
ion (Cu_2^+)

H^+ H^+

H   O
   H

copper metal   copper oxide

water molecule

lemon juice, that contains citric acid. Acids release charged hydrogen atoms known as hydrogen ions (H^+), that can react with the negatively charged oxygen in the copper oxide and turn it into water (H_2O). The copper that was associated with the oxygen then dissolves, leaving behind a pristine copper metal surface.

If you soak a large number of coins in colorless vinegar you will see the effects of the dissolved copper because the vinegar will develop a greenish tinge. This is copper acetate, and if you evaporate all of the vinegar away by placing the liquid on a saucer near a hot surface you will be left with some pretty green crystals—but don't eat them!

HOW does this apply to the real world?

Just as copper coins react with oxygen in the air, iron in the bodywork of cars, trucks, and ships will also oxidize if it gets the chance. The result is iron oxide, or rust. Luckily, the same trick that cleaned up your copper coins can be used to solve the problem. When a car develops a rust spot, mechanics first remove the paint and loose material and then treat the problem area with concentrated phosphoric acid. This reacts with the rust, stripping away the oxide and replacing it with a layer of iron phosphate. As well as treating

the rust, this protects the underlying metal against further corrosion. The excess phosphoric acid is then wiped away, and the area can be repainted. Incidentally, phosphoric acid is also a major ingredient in cola; that is part of the reason why it so effectively dissolves teeth!

SOME OTHER THINGS TO TRY

Try other foods and chemicals in the kitchen to see if any of these produce the same effect. In particular, have a go with lemonade or a soft drink.

MAKE YOUR OWN WATER STRIDER/POND SKATER

Why are bubbles round, and why don't pond skating insects sink? In this experiment we'll reveal the power of the surface tension of water.

- ◆ You will need a large, clean, dry bowl, water, matchsticks, and some detergent.

- ◆ Fill the bowl with water and allow it to settle.

- ◆ Gently float the matchstick onto the still surface of the water.

- ◆ Put some detergent on your finger and touch the water on one side of the matchstick. Instantly, it will zip across the surface, away from the side of the bowl to which you added the detergent.

WHY does it work?

It's all down to surface tension, an intellectual leap made in the 1900s by a German housewife and amateur scientist, Agnes Pockels, while she was washing up. Water molecules, which have the chemical formula H_2O, are shaped like miniature boomerangs with two hydrogen atoms forming the "arms" linked by an oxygen atom sitting at the apex. The hydrogen atoms are slightly positive, while the oxygen is slightly negative and, since unlike charges attract, the hydrogen from one water molecule is attracted to the oxygen of an adjacent molecule. This is known as "hydrogen bonding" and it glues the molecules together, making water inherently sticky.

As a result, water molecules within a body of liquid are completely surrounded by other water molecules to which they are linked by hydrogen bonds. But molecules at the surface cannot form bonds in all directions because there is nothing above them apart from air. This means that moving a molecule to the surface from within a body of liquid consumes energy, because more bonds have to break than are reformed. To combat this, the surface is always trying to shrink in order to minimize the number of broken bonds, and this shrinking force is the surface tension.

So why did the matchstick zip away across the water when some soap was added? It happened because all detergents are surfactants. They form a layer above the liquid to which water molecules can bond, meaning that the surface no longer tries to

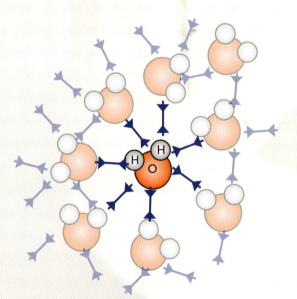

Molecules in the water body

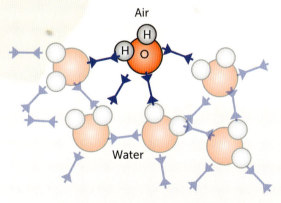

Molecules at the water surface

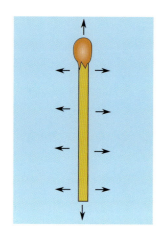

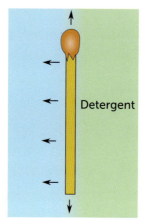

Detergent

shrink because no bonds have to be broken for molecules to move there. This has the effect of breaking down the surface tension.

As the soap was added on one side of the matchstick only, the surface tension on the opposite side remains intact, at least for a while, and continues to pull on the matchstick. Where previously the matchstick was being pulled equally in all directions and did not move, the loss of the surface tension on one side creates a net force pulling the matchstick in just one direction, that drags it across the bowl.

HOW does this apply to the real world?

Surface tension is the property that enables you to blow round bubbles, causes the rain landing on your newly waxed car to form round drops, and prevents pond skating insects from drowning.

In the case of bubbles and raindrops, the sphere is the most effective shape into which to pack the most water with the least surface area. The aim is to link the maximum number of water molecules to other water molecules, which is the most stable configuration. So surface tension pulls the water molecules into a round droplet.

Water striders/pond skaters use a slightly different strategy. If you watch one, you can see that each of its legs creates a shallow "dip" where it touches the surface of the water.

This dip increases the surface area of the water, so the water responds by trying to reduce its area again, and, in the process, pushes upward on the insect, supporting it.

SOME OTHER THINGS TO TRY

You can reveal the surface tension for yourself by floating a needle or a paperclip on water. Dropped straight into water, both of these objects would sink. But if you first lay them on top of a small piece of toilet paper or newspaper and float this onto a water surface, the paper will sink, leaving the needle or paperclip suspended by the water's surface tension.

MAKE YOUR OWN MAGNIFYING GLASS

Mention magnifying glasses and, faster than you can say Sherlock Holmes, most people are summoning up images of detectives armed with huge glass hand lenses. But you can make a lens from many things, including, believe it or not, water.

◆ For this experiment, you will need an empty plastic drink bottle, a pair of scissors, a piece of white paper, some water, and a bright ceiling light.

◆ Carefully use the scissors to cut a circular section, about 3 inches (8 centimeters) in diameter, from the top of the bottle where it narrows toward the neck. You should end up with a curved dish-shaped piece of plastic.

◆ Half-fill this dish with water and hold it above your piece of white paper and below the bright ceiling light. You might want to use a piece of cardboard with a hole cut in it to steady your lens. You should see an image of the light on the paper "screen."

◆ Move the dish up and down. You should be able to focus the image and make it fairly sharp. The water is behaving like a convex lens, exactly as you would find in a camera, a magnifying glass, or a pair of reading glasses.

WHY does it work?

It's due to the same reason that your toothbrush looks bent when you hold the handle under water. This is the process of refraction that was discovered in the 1600s by the Dutch mathematician and astronomer Willebrord Snellius.

When light passes from the air into the water in the dish, which is a denser medium, it slows down. When it leaves the water and re-enters the air beneath the dish, it speeds up again. Since the lower surface of the water is curved (because it takes on the shape of the plastic dish), the side of a light ray closest to the edge of the lens will re-enter the air slightly sooner than the edge of the ray closer to the middle of the lens. This has the effect of bending the path of the light toward the center of the lens, focusing the light toward a point.

A good way to think of this is to imagine a tractor moving along a driveway. At the end of this driveway is a small grass lawn that is shaped like the lens in this experiment. Beyond the grass is a further driveway. The left-hand wheel of the tractor drives through the middle of the lawn, while the right-hand wheel drives across the right-hand side. Like the light passing through the lens, when the tractor moves onto the grass, it slows down (perhaps its wheels sink slightly), and when it leaves the grass, it speeds up again. The right-hand wheel leaves the grass and returns to the firm driveway sooner than the left-hand wheel, because the grass is lens

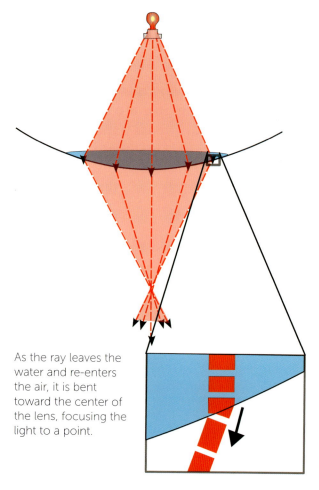

As the ray leaves the water and re-enters the air, it is bent toward the center of the lens, focusing the light to a point.

shaped. This means that when the right-hand wheel is trying to speed up again, the left-hand wheel is still moving slowly through the grass. The effect is for the tractor to be turned toward the center. This is exactly what happens to light when it passes through the water. As a result, light from above is focused to a point on the paper below. By altering the height of the "lens," you can alter its focus to obtain a sharp image of your light source.

HOW does this apply to the real world?

This is similar to the way in which a camera works. Lenses in the front of the camera focus light coming from further away to a point inside the camera where a light-sensitive microchip is mounted. When you manually focus a camera to achieve a sharp image, you are moving the lenses backward and forward in the same way that you raised and lowered your water lens to focus it in this experiment.

Lens systems like those found in cameras tend to be quite large and also involve moving parts, which are expensive and can be unreliable. Researchers are now developing tiny lenses consisting of a single droplet of liquid which can be focused by applying small amounts of electricity. These could be used to build replacement eyes for people who have lost their sight or as part of a microscopic camera system. Either way, they are effectively a miniature version of the lens you have made here.

SOME OTHER THINGS TO TRY

Add a second object at a different height to the first. Show how you can focus between them by moving the lens up and down.

Test the magnifying powers of your lens by placing objects beneath it. How much larger do they appear?

HOMEMADE LAVA LAMP

At one time, it seemed like every home had one. Living rooms were bathed in the glow of rising and falling bubbles of red, blue, and green liquids. How did they work? In this experiment we'll show you how to build your own imitation of Craven Walker's famous 1960s invention.

- ◆ You will need a large, tall jar or a tumbler, some cooking oil, some water, some bright food coloring (this is not essential), and some effervescent tablets (such as fizzy vitamin-C tablets or Alka-Seltzer™).

- ◆ Add water to the jar or tumbler until it is about one-quarter full and stir in food coloring until the water is darkly colored.

- ◆ Then add cooking oil to a depth of about three-quarters of the way up the jar. The oil will sit on top of the colored water.

- ◆ Now, drop in an effervescent tablet.

- ◆ The tablet will sink to the water layer at the bottom and produce colored foam that rises to the top of the oil and then sinks again.

WHY does it work?

This experiment is all about density, convection, and immiscibility: liquids that won't mix.

Oil floats on top of water because it is less dense and because water molecules stick to each other very strongly due to a process called hydrogen bonding. This makes it hard for any oil to get between the water molecules so the two liquids remain separated: they're immiscible. The effervescent tablet is more dense than both the oil and the water. So when it's first added, it sinks straight to the bottom.

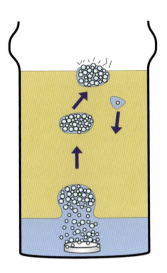

Tablets like these create fizz by combining an acid (usually citric acid) with sodium bicarbonate. When mixed with water they react, producing carbon dioxide that makes the gas bubbles that you see rising in the liquid.

The gas bubbles are much lighter (less dense) than both the water and the oil, so they form a foam that rises through both layers. When they reach the surface, the carbon dioxide escapes into the air, and the residual water carried up with the bubbles sinks back to the bottom of the jar.

HOW does this apply to the real world?

A real lava lamp uses heat rather than carbon dioxide to create the effect, but the principle is almost identical. At the base is an electric light bulb that heats a flask filled with water. The flask also contains a blob of a waxy material that, when it's cold, is slightly denser than the water, so it sits at the bottom.

As the wax warms up it melts and expands until, eventually, it becomes less dense than the water and it rises to the top of the flask.

At the top, a long way from the heat of the lamp, the temperature is much lower, so the wax cools and begins to shrink. This makes it denser than the water again, so it sinks back to the bottom of the flask, and the process starts again.

This is known as convection. The same process causes hot magma to rise toward the Earth's surface from deep within the planet, carries smoke up a chimney, and causes warm air to rise into the sky triggering a breeze as cooler air rushes in from elsewhere to replace it.

SOME OTHER THINGS TO TRY

You can create a similar effect by preparing an oil and water mixture—containing three parts oil to one part water—and then sprinkling in salt or sugar. As the crystals sink, they pick up a coating of oil. At the bottom of the glass the oil is released and rises up through the water as a droplet.

STRANGE GLOWS FROM SUGAR

Have you ever noticed a strange blue flash when you peel up the flap of some self-sealing envelopes? This is triboluminescence, and in this experiment we'll shed some light on how it happens.

- ◆ You will need some lumps of sugar in a bowl, a pair of pliers, and a very dark room.
- ◆ Before you start, gather everything you need and place the items in front of you in positions that you can easily remember.
- ◆ Switch off the light and remain in the dark for several minutes to allow your eyes to adjust to the darkness.
- ◆ Without turning on any lights, pick up the pliers and a lump of sugar.
- ◆ Place the sugar in the jaws of the pliers and crush the sugar as hard as you can.
- ◆ Watch carefully as you are squeezing the pliers: you should see flashes of blue-green light coming from the sugar.

WHY does it work?

This effect is known as triboluminescence. The word comes from the Greek *tribein*, to rub, and the Latin *lumen*, meaning light. One of the first people to describe it was the English writer Francis Bacon, in 1620. He wrote "It is well known that all sugar, whether candied or plain, if it be hard, will sparkle when broken or scraped in the dark." And now we know why.

Some sugar crystals are asymmetrical. When they're crushed, they can fracture in such a way that the resulting fragments contain unequal numbers of positive and negative charges. As you continue to crush the sugar, some of these unequally charged pieces will be pulled apart. And since positive and negative charges attract, pulling them apart against this force gives the charges more energy, in the form of a high voltage.

Eventually, a point is reached where the voltage becomes large enough to overcome

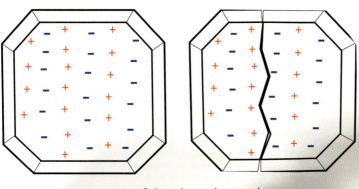

the natural resistance of the air, and a spark jumps across the fracture like a miniature bolt of lightning. The flash of light occurs because the flow of electricity excites air molecules, and this temporarily kicks electrons belonging to some of the gas molecules into higher energy states. When the excited electrons then return to their normal states, they release the energy again as light, some of which we can see.

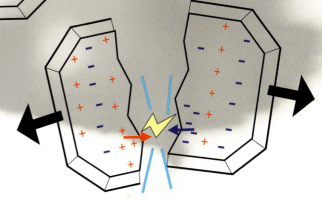

HOW does this apply to the real world?

You can see triboluminescent effects when cutting and polishing diamonds (these produce a blue or red glow), but if you can't afford a diamond then a piece of flint will do! If you throw a pebble at the ground on a beach covered with flint pebbles when it is dark, you get little flashes of green light due to triboluminescence in the silica of the flint. (Please note, wives and girlfriends will not be satisfied with the substitution of a piece of flint for a diamond on other occasions!)

SOME OTHER THINGS TO TRY

You can also demonstrate triboluminescence with some self-sealing envelopes. If you pull up the glued flap in a dark room, you will see a pale-blue flash as the glue molecules behave like your lumps of sugar and separate charges unevenly as you pull them apart.

THE MYSTERIOUS SOUND OF AN OVEN SHELF

How can an oven shelf sound like London's famous Big Ben? The answer is that it already does, if only you could hear the low frequencies it emits. Now, thanks to this experiment, you can!

- You will need the metal shelf from an oven (make sure it's cold first), some string, a wooden spoon or something similar to use as a drumstick, and an assistant.

- Cut two lengths of string, each about 1 yard (1 meter) in length, and tie each piece to adjacent corners of the oven shelf.

- Wrap the free ends of each piece of string around the index fingers of both hands and stand up.

- Hold the oven shelf up by the string so that it is dangling in midair.

- Ask your assistant to strike the metal with the spoon and listen to the sound it produces.

- Put the fingers with the string wrapped around them into your ears. You may need to lean forward slightly to make sure that the shelf is still dangling in midair.

- Ask your assistant to hit the shelf again.

- It will sound like a giant bell resonating inside your head. But to your assistant, it will sound the same as before, even if they stick their fingers in their ears.

WHY does it work?

This experiment is all about sound waves and how they travel through air and other materials. When you hit the oven shelf, the metal vibrates and transfers these vibrations to the surrounding air molecules that carry the sound energy to your eardrums. The eardrums then feed the movements into the inner ear, known as the cochlea, where they are converted into electrical nerve signals that the brain can understand.

High-frequency (high-pitched) sounds are transmitted to the air much more effectively than lower frequencies, which is why, with the oven shelf hanging freely, you hear just high-pitched ringing tones.

This is because air is fluid, and quite similar to water. If you wave a hand through water, you feel very little resistance because the water can easily flow around your fingers. You will also make very few waves. But if you slap the water, or try to move fast, it feels much stiffer because the water doesn't have time to move out of the way. Instead, it piles up in front of your hand and forms waves.

In the same way, the high-frequency vibrations from the oven shelf force the air to move quickly, so it piles up and forms sound waves that you can hear.

However, the low frequencies vibrate the air much more slowly, giving it more time to move aside, so the resulting sound waves are much weaker and harder to hear.

At high pitches, the rapid vibrations mean that the air doesn't have enough time to move out of the way and so it piles up, forming waves.

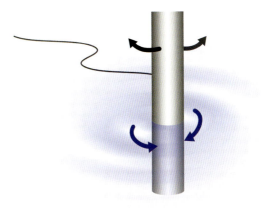

At low pitches the air has time to flow around the slowly vibrating oven shelf, so only weak waves are produced.

What does the string do? Supporting the weight of the oven shelf makes the string stiff, so it behaves as an ideal conduit for the slower, low-pitch sound waves that pass up the string and into your fingers.

By sticking your fingers in your ears, you make contact with the bones of your skull that carry the low-pitched sound waves directly to the cochlea in your inner ear where they are converted into nerve signals. Now you hear the low as well as the high frequencies.

HOW does this apply to the real world?

This experiment shows you why your voice appears to sound different when you hear a recording of yourself. When you speak, the sound reaches your cochlea in two ways. Some of the sound waves leave your mouth and travel around your head to your eardrums, where they are picked up the normal way, but the vibrations inside your mouth are also transmitted to the bones of your skull. These carry them directly to the cochlea without passing through the air.

What you hear when you speak is a combination of both the bone and air conduction of the sounds you make. This is like listening to the oven shelf with your fingers in your ears. But when you hear a recording of yourself, the sounds arrive only via the air, so you hear a different set of frequencies. This is like the oven shelf hanging freely. Now you know why the recording you made of yourself the other day doesn't seem to sound like you at all. Worse still, that sound is what everyone else hears!

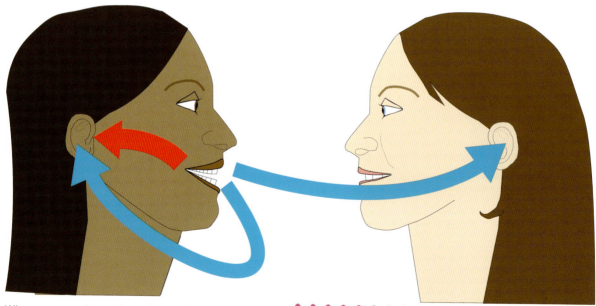

When you speak, you hear the combined effects of sound waves conducted through the air and the bones of your skull, but the person you are talking to hears only the sounds that have traveled through the air. So, to them, you sound different.

SOME OTHER THINGS TO TRY
Experiment with other objects to find out whether there are other things around the house that have the same effect.

JUST CHILL OUT...
YOUR GLOW-STICK

Many of the key processes we see in the world around us are powered by chemical reactions. The human body turns dinner into energy; power stations turn coal into electricity; and plants turn carbon dioxide and water into sugars and oxygen. What are chemical reactions, and how are they controlled?

◆ For this experiment, you will need some glow-sticks (like the ones sold at fairgrounds), some ice, a bowl, and some salt.

◆ Begin by adding the ice to the bowl, and pour in water to a depth three-quarters of the height of the ice.

◆ Sprinkle several tablespoonfuls of salt over the ice surface. This will reduce the temperature by as much as 32°F (18°C).

◆ Then take a glow-stick (or a few if you are feeling extravagant) and bend or crush it so that it begins to give out light. Shake it to make sure that the chemicals are well mixed.

◆ Now poke the glow-stick into the ice, leaving half of it exposed above the surface of the ice. Alternatively, if you don't have any ice, you could use two glow-sticks and place one in a freezer and keep the other at room temperature.

◆ After ten minutes, remove the glow-stick from the ice and compare the two ends. One should be glowing more brightly than the other.

◆ Warm the cold end of the glow-stick by rolling it between your palms for 20–30 seconds and compare the two ends again.

◆ Depending upon how cold you make it, the chilled part of the glow-stick can almost stop giving out light altogether. Then, when it warms up again, its brightness returns.

WHY does it work?

This is all about reaction kinetics and explains why animals that are active all year round, like humans, are also warm-blooded. Glow-sticks use a chemical reaction to produce light. They consist of two tubes, one inside the other, each containing different chemicals. The innermost tube is very fragile and filled with hydrogen peroxide (that is often used as hair bleach). When the glow-stick is bent or squashed, this tube breaks, and the hydrogen peroxide escapes into the outer tube where it mixes with a second chemical, diphenyloxalate ester, and a fluorescent dye.

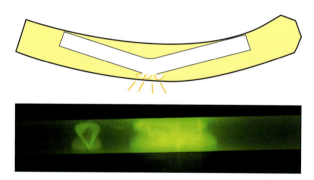

The hydrogen peroxide reacts with the diphenyloxalate ester; this breaks down, releasing energy as it does so. The energy is picked up by the dye molecules that become excited and re-release it as light. One of the other products of the reaction is carbon dioxide, which is why spent glow-sticks often contain small bubbles.

Chemical reactions like these rely on collisions occurring between the reagents (the hydrogen peroxide and the diphenyloxalate ester particles). If they hit each other hard enough then they will react. As anyone who's ever bumped into something knows, the faster you are moving, the more it hurts, because the energy involved in the collision is greater.

The same is true with chemical reactions. The higher the temperature the faster the particles move and, hence, the more numerous and the

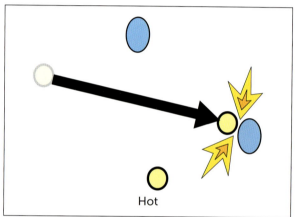

Hot

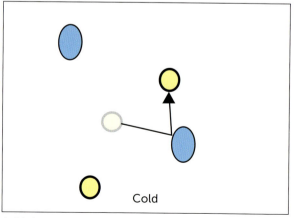

Cold

At higher temperatures, particles have more energy so they hit each other harder when they collide, making a chemical reaction more likely.

more powerful the collisions between them. As a general rule, the rate of a chemical reaction approximately doubles for every 18°F (10°C) increase in temperature.

The reverse is also true. If, instead, things cool down, the particles move more slowly, collide with each other less often and less hard, and the rate of the chemical reaction drops. Since this chemical reaction makes light, the slower the reaction, the dimmer the glow-stick becomes, so the half of the glow-stick cooled in the ice produces less light.

HOW does this apply to the real world?

Some animals, like lizards and frogs, are cold-blooded, meaning that they cannot generate their own body heat like a dog or a human. The temperatures of their bodies vary with the temperature of the environment, so when it becomes cold, their body temperatures drop, and the chemical reactions that power metabolism also slow down. This can affect how energetic these animals are and can limit how quickly they're able to move. Sluggish lizards will find it more difficult to catch food or to escape from predators so, to overcome this problem, they bask in the sun. By turning themselves into living solar panels, they soak up heat, which boosts body temperature and increases metabolic rate, ensuring they can make a fast getaway when they need to!

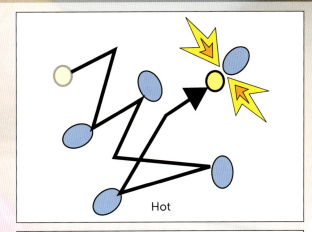

Hot

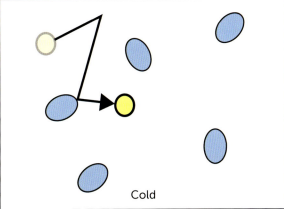

Cold

Hot particles are moving faster so they will collide with other particles more often. This also increases the likelihood of a chemical reaction.

SOME OTHER THINGS TO TRY

If you have an active glow-stick one evening and would like to keep it overnight, put it in the freezer. To reactivate it, warm it up again!

TANGTASTIC: MAKE YOUR OWN FIZZY CANDY

How do those sweets work that seem to fizz when you put them in your mouth? In this sweet-toothed experiment we'll show you how to make some of your own.

- You will need some bicarbonate of soda (baking soda), some citric acid, some confectioners' sugar (or superfine sugar/icing sugar), and a dry bowl in which to mix them. If you wish to keep your fizzy food, you will need a container with a tight lid to keep out moisture.

- Mix 1 teaspoonful of bicarbonate with 3 teaspoonfuls of citric acid, and then stir in 7 teaspoonfuls of confectioners' sugar/icing sugar.

- Now take a small amount on a spoon and eat it!

- As soon as your saliva touches the powder, it should begin to fizz (you may even hear it), and your mouth will fill with bubbles.

WHY does it work?

You have unleashed on your taste buds a chemical reaction between an acid and an alkali. One of the products is carbon dioxide; this is the fizz you feel on your tongue. Acids are chemicals that, when added to water, release charged particles called hydrogen ions. These are highly reactive and will combine with certain other chemicals in order to become more stable.

One substance that they will readily attack is bicarbonate; this contains carbon dioxide linked to an extra atom of oxygen and hydrogen. When a hydrogen ion reacts with a bicarbonate, it releases the carbon dioxide and joins with the

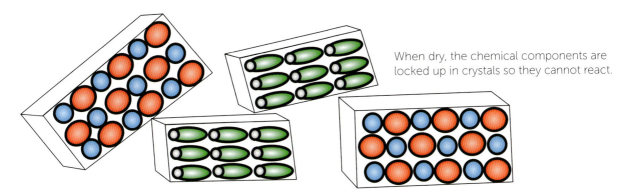

When dry, the chemical components are locked up in crystals so they cannot react.

extra oxygen and hydrogen to make a molecule of water. Both of these products are very stable, which is why the reaction takes place.

The acid, in this case citric acid, can only produce the hydrogen ions that attack the bicarbonate if water is present. This is why the citric acid and bicarbonate can be mixed together in a dish but will only begin to react when they get wet, such as in your mouth. Water also enables the acid and the sodium bicarbonate to dissolve. This makes it easier for the two chemicals to mix.

When dissolved in water, the components can mix and react.

HOW does this apply to the real world?

As well as in sweets, this type of reaction is also used to produce effervescent pills and tablets for treating headaches and fizzy vitamin-C preparations.

It's also at work in the kitchen. Baking powder and self-raising flour both contain a mixture of an acid (such as tartaric acid) and sodium bicarbonate. When water is added, the two chemicals begin to react, producing carbon dioxide just like in the experiment

above. The gas becomes trapped inside the food mixture in the form of tiny bubbles that inflate your cake and make it rise.

SOME OTHER THINGS TO TRY

Demonstrate this baking powder reaction by adding a teaspoonful to a small amount of hot water. You should see it fizzing. Don't confuse baking powder with baking soda: this is just the sodium bicarbonate without any added acid, and getting them confused could lead to some very flat cakes.

PLANT HYDRAULICS, AND WHY SLUGS AND SALT DON'T MIX

What makes plants wilt on a hot day but then stand up straight again within minutes of being watered? In this experiment, with the help of a potato, we'll reveal the power of the wet stuff and explain why slugs shrivel up at the sight of salt.

- You will need a reasonable-sized potato, a knife and chopping board, some salt (or sugar), hot water, a jug, and two bowls.

- Begin by adding about 1-2 cups (300 milliliters) of hot water to the jug, and then dissolve in it as much salt (or sugar) as you can. The solution is ready to use when no more will dissolve.

- At this point, place the jug to one side to let it cool.

- Peel the potato and cut it into long "french fries." Aim to make them at least 2 inches (5 centimeters) long by about ½ inch wide (1 centimeter) and ¼ inch thick (.5 centimeters).

- Notice that the pieces are rigid, and if you hold them by one end they remain straight.

- When the salt or sugar solution in the jug has cooled, pour the liquid into a cereal bowl.

2

- Add several of your french fries to the bowl, making sure that they are covered by the liquid.
- Then fill a second cereal bowl with fresh (tap) water and add a similar number of fries.
- Make sure that you know which bowl is which and leave them to soak overnight.
- In the morning, remove a fry from the bowl containing fresh water. It should feel stiff, and held just at one end it should remain straight. If you try to bend the two ends toward each other, the fry should snap.
- Now take one from the salt or sugar solution. It will have become completely floppy and you may even be able to bend it into a circle!

WHY does it work?

This is all about the science of osmosis, first described in the early 1800s by the French biologist Henri Dutrochet. Just like our bodies, plants are composed of millions of tiny cells, each one less than a thousandth of an inch (a fiftieth of a millimeter) across. The cells are surrounded by cell walls made from a tough polymer called cellulose, that helps them to keep their shape. Inside each cell wall is an oily layer called the cell membrane. This acts like a sieve to control what substances enter and leave the cell. Water (the blue molecules in the diagram) can move through the cell membrane without difficulty, but other chemicals (red), such as sugars and salts, require specialized transporters to move them into and out of the cell.

Cells usually contain large amounts of these substances dissolved inside them, so their contents are more concentrated than any liquid outside the cell. This has the effect of drawing water into the cells to dilute the contents, and it also makes the cells swell up inside the confines of their cell walls.

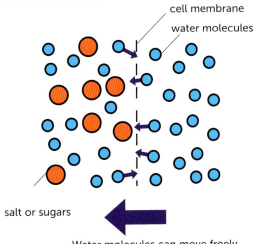

Water molecules can move freely across the cell membrane by osmosis

This movement of water to dilute a concentrated solution is called osmosis, and it continues until the cell contents are at the same concentration as the surrounding liquid, or until the cell wall pushing back in on the cell contents prevents any further water from moving in.

The water-filled cells press tightly upon their neighbors, making the plant structure very stiff and rigid, which is why a freshly cut french fry, or one that has been soaked in fresh water, doesn't bend.

Why do fries go floppy when they are left in sugar or salt solution? This is because the concentration of your salt or sugar solution is much greater than the concentration of sugars and salts inside the plant cells. As a result, water moves out of the cells to dilute the more concentrated solution in the dish.

This causes the cells to deflate, introducing empty spaces into the tissue and making it able to bend.

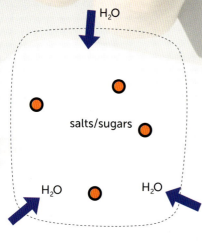

In fresh water

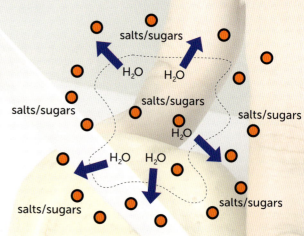

In strong salt/sugar solution

HOW does this apply to the real world?

Precisely the same process occurs when cut flowers run out of water and begin to droop, or when plants wilt on a hot day. Thankfully, if it's caught soon enough, the effect can be reversed by replenishing the water in the cells before permanent damage is done.

Osmosis is also the reason why slugs shrivel up if they are coated with salt. The crystals dissolve in the damp mucus covering the slug's body and begin to pull water out of the tissues, causing the slug to dehydrate and die.

SOME OTHER THINGS TO TRY
Swap the fries from the salt solution to the fresh water and vice versa. Can you turn a limp fry back into a stiff one?

HANG AN ICE CUBE FROM A THREAD

If you didn't have a freezer, could you still make ice cream? Yes, and this experiment will show you how. It will also reveal why we put salt on the roads in winter.

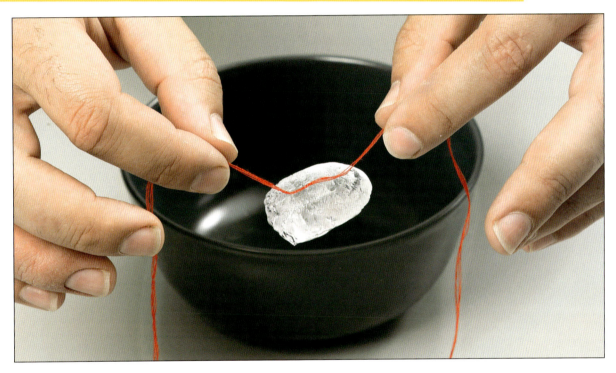

◆ You will need some ice cubes, a length of cotton thread (6 inches [15 centimeters]), a bowl, and some salt.

◆ Begin by placing a single ice cube in the bowl. The experiment works best if the cube is just beginning to melt so that the surface is covered with a thin layer of water.

◆ Gently lay the cotton across the top of the ice cube.

◆ Sprinkle a very small amount of salt onto where the thread touches the ice.

◆ After adding the salt, wait about 15 seconds, and then pick up the cotton. The ice cube will be frozen to it. But where the salt landed, the ice-cube will be wet.

WHY does it work?

This is down to the science of thermodynamics. Your ice cube is stealing energy from one part of its surface and giving it to another. When water freezes, the molecules pack together in a very stable regular shape, that is held in place by the formation of bonds between each molecule and its neighbors. This is why melting ice takes a large amount of energy, because these stable bonds have to be broken to free the water molecules again.

Solid as this icy arrangement sounds, if you could watch the individual molecules on an ice surface you would see some of them detaching from the ice and becoming a liquid again, while others would be refreezing into the crystal. If you don't heat or cool the ice, the rates of melting and freezing are in balance or equilibrium, with equal numbers of water molecules joining and leaving the ice crystal.

When salt is added, it disturbs this equilibrium by preventing the water from refreezing. It gets between the molecules and makes it harder for them to find their way back to the ice surface, so instead they remain as a liquid. The result is that more bonds are being broken in the ice, to release liquid water, than are being formed by water molecules refreezing. The energy to break these bonds has to come from the ice cube itself, and, as it loses this energy, its temperature drops and can reach as low as 0°F (-18°C).

Why does the cotton stick to the ice? This is because the salt isn't uniformly spread over the surface of the cube, so there will be salt-free patches of ice, particularly beneath the cotton. As the ice cube cools down due to the melting effect of the salt, the pure water in the salt-free areas, including inside the cotton, immediately freezes and glues the thread to the ice.

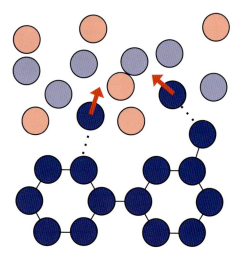

Without salt, water molecules join and leave the ice surface.

When salt (red) is added, water molecules (blue) find it harder to return to the frozen surface, so the ice progressively melts.

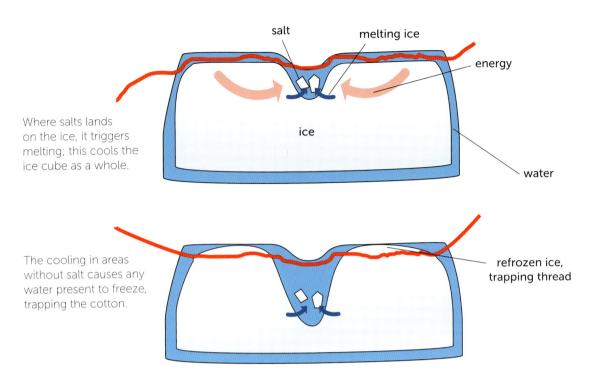

Where salts lands on the ice, it triggers melting; this cools the ice cube as a whole.

salt

melting ice

energy

ice

water

The cooling in areas without salt causes any water present to freeze, trapping the cotton.

refrozen ice, trapping thread

HOW does this apply to the real world?

Before anyone had freezers they still had ice cream, and it was made using precisely this method. Ice (collected during the winter and stored underground in "ice-houses") was crushed up and mixed with salt in a large bowl. Its temperature would instantly plummet, as you saw in this experiment. This frosty mixture was then used to rapidly freeze the contents of a second bowl, placed inside the first and containing the cream and sugar. Thankfully, these days we have freezers!

The science behind this experiment also helps to keep roads frost-free in winter. When bad weather is expected, rock salt is scattered over the road surface. This lowers the temperature at which ice will form so that any water on the road remains as a liquid, preventing the road from turning into an ice-rink. However, if the temperature drops low enough, freezing can still occur, and a different form of salt is needed. An environmentally friendly, albeit more expensive, choice is sodium acetate, the same chemical used to flavor salt and vinegar chips/crisps!

SOME OTHER THINGS TO TRY

Try making your own "ice cream" using a salt and ice mixture to cool down a small bowl of mixed cream and flavoring, such as vanilla essence.

LIQUID BEHAVING BADLY

Imagine a substance that you could not only walk on but also swim in or roll into a firm ball and then watch melt into a liquid in front of your eyes. It sounds like science fiction, but in all likelihood the ingredients to make this mysterious material are sitting in your kitchen cupboard.

1

2

3

- For this experiment, you will need a packet of cornstarch/cornflour, water, and a large bowl in which to mix them together.

- Pour the cornstarch/cornflour into the bowl.

- Add a small amount of water and mix the two with your hands.

- Keep adding water in small amounts. As a guide, a 16-ounce (500-gram) box will need about a cup (300 milliliters) of water.

- When the mixture is right you will have a white runny liquid that behaves like a solid when you push it, hit it, or roll it into a ball but turns back into a liquid when you stop moving it.

- If you were to fill a bath with your mixture you could walk on it, although you would sink if you stood still!

WHY does it work?

The wet cornstarch/cornflour is behaving as a "shear-thickening" fluid. It's effectively the reverse of quicksand, and the same science is involved in the safe removal of oil from the ground and in making body armor that is comfortable to wear but that gets tough when it needs to be.

Under a microscope, cornstarch/cornflour consists of billions of tiny irregularly shaped particles of starch, each less than one thousandth of an inch (one hundredth of a millimeter) across. When water is added, the liquid flows around each starch grain and acts like a lubricant, making the mixture runny by helping the particles to slip over each other.

When a sudden large force is applied, the starch grains jam together, squeezing some of the water out from between them.

Without the lubricating effects of the liquid, the particles cannot slide past each other and so the mixture starts to behave like a solid; it can be rolled into a ball, bounced, fractured, and snapped. However, these effects are only temporary. As soon as the force is removed and the mixture is allowed to "relax," the water resurrounds each of the particles and it becomes runny again.

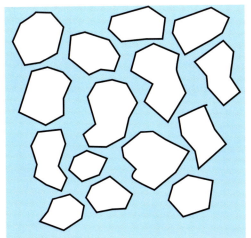

The starch grains are surrounded by water, that allows them to flow past each other, like a liquid.

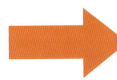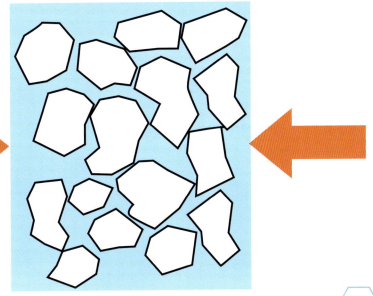

Applying force to the mixture squeezes out water from between the particles, jamming them together so they behave as a solid.

HOW does this apply to the real world?

Shear thickening can be a serious problem during oil exploration when a borehole is drilled into a reserve. The drill is continuously cooled and lubricated by pumping through it a substance called "mud," that also removes fragments of rock as they are cut away. The mud returns to the surface, where it is cleaned and recycled, but drilling engineers have to monitor very closely how much rock it contains. Too much, and the fragments could jam together like the cornstarch/cornflour in this experiment. A pump moving the lubricant would find itself trying to compress a substance with the consistency of solid rock, probably with expensive if not explosive consequences! Scientists also analyze the composition of the particles in the mud to look for geological clues to the presence of an oil field below.

Shear-thickening fluids can also be extremely useful, and engineers are now exploring their use in body armor, such as bulletproof vests. These are usually bulky. They restrict movement and only protect the torso, leaving the wearer's legs and arms exposed. To get around the problem, scientists have come up with a way to suspend tiny silica (glass) particles in a non-toxic fluid. When pressed lightly, the mixture flows easily, but if a bullet or a piece of shrapnel strikes it, the large force causes the fluid to instantly behave like the cornstarch/cornflour in this experiment. The silica particles jam together and solidify, fending off the projectile and protecting the wearer. Because it doesn't need to be as thick as traditional body armor and is also very flexible, the limbs can be protected too.

SOME OTHER THINGS TO TRY

Wriggle your toes in the damp sand at a beach—the principle is the same. Usually, large particles of sand squash water out from between them under the weight of a person. But wriggling your toes helps particles become suspended and flow past each other, so you begin to sink. (The effect can be very dramatic—whole buildings have sunk in this way, when soil has "liquified" during an earthquake.)

INVISIBILITY CLOAK

Since glass is transparent, why can we see it at all? Why is it not invisible? Well, it can be, and in this experiment we'll show you how to make it vanish.

- ◆ You will need a large glass bowl, a bottle of (cheap) cooking oil, and a smaller Pyrex™ glass bowl that will fit inside the larger bowl.
- ◆ Start by pouring the cooking oil into the larger glass bowl.
- ◆ Pick up your Pyrex™ dish and immerse it in the cooking oil.
- ◆ Look from the side. Can you see it?

WHY does it work?

This experiment is all about refraction and how light changes its speed when it passes from one substance (or medium) into another.

Light travels through different substances at different speeds. It moves fastest in a vacuum, slightly slower in air, and slower still in water or glass. When it leaves one medium and enters another—such as passing from air into water—it has to alter its speed, and this also causes the path of the light to bend. This is the reason why we can see the glass bowl under normal circumstances, because light passing from the air into the glass slows down and alters its direction. This distorts the view of other objects coming through the glass, that, along with some reflections from the glass surface, tell the brain that there must be a transparent object in the way. As a result, you can see transparent glass.

So why does the bowl disappear when you immerse it in cooking oil? This is because, by chance, light travels at almost the same speed in cooking oil as it does in Pyrex™ glass. As a result, when light passes from one into the other it doesn't alter its speed and therefore it doesn't change direction. Your brain has no way of knowing that the light coming through the oil has also passed through the Pyrex™. With nothing to distinguish where one ends and the other begins, you cannot work out where the bowl is, so it disappears!

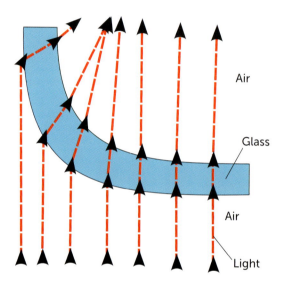

Air

Glass

Air

Light

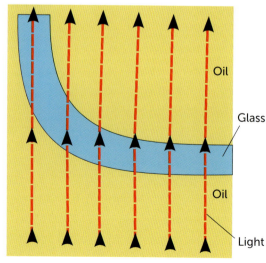

Oil

Glass

Oil

Light

HOW does this apply to the real world?

This is the reason why it is easy to cut yourself on broken glass in the sink, and why glasses look shiny when they're wet.

Although it's not exactly the same, the speed at which light travels in glass and in water is sufficiently similar that light isn't refracted very much when it passes between them. So if you break a glass while you are washing up it can be very hard to see the broken shards beneath the water.

This is also why an old drinking glass looks shiny and new when it is wet, but frosted and scratched when it dries. Scratches on the glass surface cause light to reflect and refract in different directions compared with the adjacent unscathed glass, and this makes the scratched parts look white. However, when the glass is wet, the scratches are filled and smoothed out by a layer of water, so they disappear, a bit like the Pyrex™ bowl in this experiment.

SOME OTHER THINGS TO TRY
Try the experiment with non-Pyrex™ glass—does it work as well?

TURN MILK INTO CHEESE

What is milk and why does it go all lumpy if you leave it out of the refrigerator too long? In this experiment, we'll show you how to separate milk into some of its component parts the "ch-easy" way.

◆ For this experiment you will need some milk, a glass, some vinegar (or lemon juice), and a coffee filter or some kitchen paper towel.

◆ Add some milk to the glass.

◆ Add small amounts of vinegar, swirling the mixture as you go.

◆ Quite soon you should see the milk forming solid white flakes that will sink to the bottom of the glass.

◆ Swirl the glass to re-suspend the particles and strain the mixture through the coffee filter or the paper towel.

◆ You will be left with a white solid material that can be pressed into a firm lump. This is a mixture of fat and the protein casein. Casein makes up the bulk of milk solids and is the material used to make cheese. The liquid you strained off should be clear: this is the "whey," a solution of the milk sugar lactose and minerals including calcium and phosphorus (plus some vinegar of course).

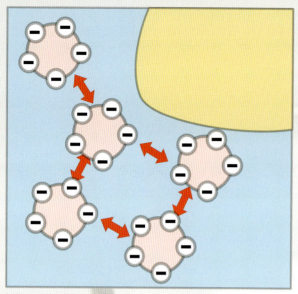

Milk contains a mixture of water (blue), fat globules (yellow), and casein micelles (pink), that are kept apart by their negative charges.

Positively charged hydogen ions neutralize the negatively charged casein.

WHY does it work?

It's because vinegar is an acid and it can trigger proteins like casein to change their shapes and solubility. This process is known as denaturation, and heat and alkalis can do it too. Proteins like casein are biological polymers that are made by linking together chemical building blocks called amino acids. Some amino acids tend to repel water, while others are attracted to water, so the protein arranges itself into a shape that puts the water-loving amino acids on the outside of the protein and the water-repelling amino acids together on the inside. This behavior is what gives the protein its structure and its characteristic behavior.

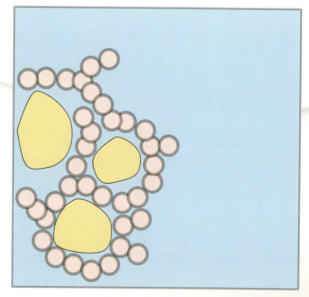

The neutralized casein micelles link up, encasing the fat globules. The watery whey is left behind.

In milk, groups of casein proteins cluster together to form tiny ball-shaped structures called casein micelles. They each measure less than one hundred-thousandth of an inch (one-ten-thousandth of a millimeter) across, which is so small that they remain suspended in the surrounding water. The amino acids on the outsides of the proteins also carry a negative charge that keeps individual micelles apart and stops them from sticking together.

Acids, in the form of vinegar or lemon juice, contain hydrogen ions. These carry a positive charge and surround the casein micelles, neutralizing their negative charge and preventing them from repelling one another. The casein micelles then begin to link together, forming large clumps that are the white solids that appeared in the solution.

HOW does this apply to the real world?

You can see the same chemical process at work in your own refrigerator. When a bottle of milk "goes bad," it will separate into a pale watery layer at the top of the bottle, that is the whey, and a thick heavy white layer toward the bottom, that is the casein, or "curds." This is why milk souring is known as "curdling."

Usually this occurs because bacteria growing in the milk convert lactose sugars, that are part of the whey, into lactic acid. The lactic acid releases hydrogen ions like the vinegar in this experiment; this triggers the casein to aggregate. This is why milk stored in the refrigerator stays fresh for longer because the low temperature slows down the growth of the microbes, reducing the rate at which they produce lactic acid.

This process isn't all bad news, because it is also the key to making cheese. Milk can be curdled by adding extra bacteria to produce more acid, but usually manufacturers add an enzyme called rennet. This is produced in the stomachs of newborn calves and helps to make milk more digestible by causing the casein to stick together. Once the casein has formed solids, the whey is strained off, the excess water is squeezed out, and the cheese is left to mature. The chemical actions of residual bacteria and added fungi and molds impart characteristic textures and flavors. And the whey? It used to be fed to animals, but now the lactose sugar is removed and used in pills and tablets.

IS THAT REALLY ORANGE?

When is orange not orange? They might look the same, but not all orange-colored lights are made equal, as this experiment will reveal.

◆ For this experiment you will need a magazine filled with colorful pictures and photographs (1), a road lit by old-fashioned orange streetlights (or a red LED bicycle lamp), and the orange indicator or hazard flashers of a car (you don't need to remove them from the car!).

◆ Wait until it is dark.

◆ Take the magazine outside and look at it beneath the orange street lights. The colors will now look either different tones of orange (red if you used the bike light), or just black (2).

◆ Now hold the magazine close to the flashing orange indicator lights of a car. This time, when the light flashes on, the colors are all visible, although the reds and yellows will be much brighter than the blues and greens (3).

WHY does it work?

This effect is down to the science of spectroscopy, that was worked out by two German scientists in the mid-1800s, physicist Gustav Kirchhoff, and chemist Robert Bunsen, who also invented the famous burner that bears his name. Their discoveries allow space scientists to work out the composition of distant stars just by looking at them and the same science ensures that television presenters look the right color on-screen.

Isaac Newton first showed over 300 years ago that white light is actually a mixture of colored lights blended together. This spectrum is revealed whenever sunlight shines through rain and produces a rainbow, or when white light is directed into a prism that splits up the beam into its component colors. Each of these colors corresponds to light waves with a different wavelength, that range from the very short wavelengths of purple and blue lights to the much longer wavelengths of yellow and red.

When white light shines on an object, the surface is hit by light of many different wavelengths. Some of these will be absorbed by the surface, but others will be reflected, and this is what determines the color. For instance, an object that looks red is absorbing all the wavelengths of visible light except red; it reflects red back at you. A black object, on the other hand, absorbs all the visible light that hits it and reflects very little back, which is why it looks dark.

If just orange light is shone onto a surface, there are now only two choices: either the inks on the surface absorb orange light—and look dark—or they reflect orange light and look orange.

This monochrome effect is what you saw under the streetlights that are also known as sodium lamps. They work by passing electricity through a cloud of sodium atoms kept at a low pressure. The electrical energy causes electrons orbiting the sodium atoms to temporarily flip into higher energy levels. The electrons quickly fall back to their normal energy level, but as they do so they re-release the extra energy they absorbed in the form of light at a very specific wavelength. For sodium, this wavelength corresponds to an orange color, but other chemicals have their own specific wavelength, and this was the basis of Kirchhoff and Bunsen's discovery.

What about the car indicator? It looked orange too, yet all of the colors were visible on the page. This is because there is more than one way to make orange. Streetlights produce just orange light and very little else, but the car indicator light uses a white bulb that shines through an orange filter to produce its color. As a result, the orange light coming from the indicator contains other wavelengths of light mixed in, including reds, oranges, yellows, and a bit of green. When you illuminate the magazine using the indicator light, the extra wavelengths present are reflected by the inks on the page, revealing most of the colors.

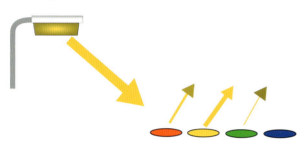
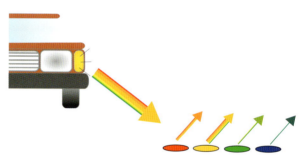

HOW does this apply to the real world?

This experiment shows you why things can look different colors under different forms of lighting. This is why it's important to take a sample of paint or material home with you to try before you redecorate, because sometimes the lights in a showroom can fool you into seeing colors that will look very different when they're in your living room.

The same applies to television presenters. The make-up artist will use colors designed to make the person look a healthy color despite the fierce white glare of studio lighting.

Under different lights there is a danger of the decorated person appearing orange, as some celebrities have found to their horror!

And the composition of distant stars? As Bunsen and Kirchhoff found, each chemical element absorbs and emits light at specific wavelengths. By looking at the light coming from an object like a distant star it's possible to spot gaps in the spectrum that correspond to the different chemicals that make up that object.

HOMEMADE MINI FIRE EXTINGUISHER

Fire extinguishers come in all shapes and sizes. Some use water, others use gas, and some are dry powder extinguishers. But how do they work?

◆ For this experiment, you will need a large tall glass, some bicarbonate of soda (baking soda), some vinegar, a candle, and some matches.

◆ Add 3–4 teaspoonfuls of bicarbonate of soda to the glass.

◆ Pour in ½–1 (1–2 centimeters) inch depth of vinegar. The mixture will fizz and froth up.

◆ While you wait for this to stop, light the candle.

◆ Carefully pick up the glass and, without pouring out the vinegar, gently tip it, from a height of about 4 inches (10 centimeters), onto the top of the candle flame. It may help you to imagine that the glass is filled with an invisible liquid sitting above the vinegar and bicarbonate.

◆ Although nothing appears to leave the glass, the candle flame will abruptly go out. Relight it: you can repeat the experiment several times (although you may need to recharge your extinguisher with more vinegar and bicarbonate).

WHY does it work?

You have produced a gas that was first discovered in 1750 by the Scottish chemist and physician Joseph Black and which is a highly effective fire extinguisher. It's carbon dioxide that blankets the candle flame and puts it out.

Sodium bicarbonate contains molecules of carbon dioxide linked to additional atoms of oxygen and hydrogen. When mixed with acids, such as vinegar or lemon juice, the bicarbonate breaks down and releases the carbon dioxide as a gas; this produces the fizzing you saw. The oxygen and hydrogen join with the acid and turn into molecules of water.

Molecules of carbon dioxide are heavier than air, so as the gas fizzes out of the solution it accumulates in a layer above the liquid and pushes the lighter air out of the top of glass. By the time the fizzing has stopped, the glass should be full of carbon dioxide.

When you tip the glass above the candle flame, the carbon dioxide begins to flow out like a liquid and falls down onto the flame. A burning candle continuously evaporates wax vapors from its wick. These mix with oxygen from the air and burn to produce water and carbon dioxide. The carbon dioxide is normally carried away by the heat rising from the candle, so fresh air (and oxygen) can enter at the base of the flame.

When carbon dioxide from the glass falls from above onto the flame, it forms an invisible blanket that cuts off the supply of oxygen and puts out the candle. You can relight the candle because the carbon dioxide quickly mixes into the surrounding air and disappears.

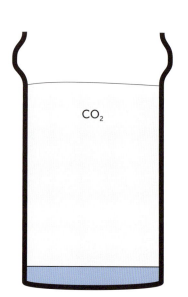

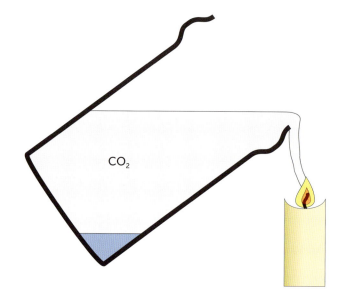

HOW does this apply to the real world?

Although most people think that water is the best way to put out a fire, there are some situations, such as in electrical or certain chemical fires, where using water could be very dangerous and even cause an explosion.

Instead carbon dioxide is used. The gas is compressed into cylinders and when sprayed at the base of a fire it cuts off the oxygen supply, putting out the fire, just like the candle in this experiment. The expanding gas coming from the extinguisher is also very cold; this can help to cool down whatever is burning to below its ignition temperature.

What about dry powder extinguishers? These rely on the same chemical you used in this experiment, sodium bicarbonate, or its relative potassium bicarbonate. When the bicarbonate is sprayed onto the fire, the heat causes it to break down, known as decomposition. This releases the carbon dioxide inside the fire and puts it out.

SOME OTHER THINGS TO TRY

You can attempt to create your own dry power fire extinguisher by using a spoon to carefully sprinkle a small amount of bicarbonate of soda onto the candle flame.

HOW TO MAKE A FORCE FIELD

Force fields and tractor beams are staples of sci-fi action space movies, but you don't need to be a rocket scientist, or even the captain of the *Starship Enterprise* to make one.

◆ For this experiment, you will need an empty metal drink can (aluminum works best because it is light), a rubber balloon, and a hairy head. (If you don't have a balloon, a piece of styrofoam can be used instead.)

◆ Begin by rubbing the balloon against your head for 20–30 seconds to charge it up with "static" electricity. The rubbing action causes the balloon to pick up electrons from your hair, giving it a net negative charge.

◆ Lay the drink can on its side on a flat, even surface and gently wave the charged balloon close to it; try not to let the two touch each other.

◆ The can will mysteriously roll toward the balloon when it comes close.

WHY does it work?

Your charged balloon creates an electric force field that spreads out around the balloon and can pull in other objects. The metal drink can contains equal numbers of positive and negative charges, so it is electrically neutral.

When the negatively charged balloon is brought close, the charged particles in the can are forced to move.

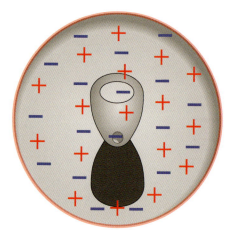

An equal number of positive and negative charges are evenly distributed throughout the can.

Since metals are good conductors and like charges repel each other, electrons in the side of the can closest to the negatively charged balloon move away toward the opposite side of the can. This leaves the side nearest to the balloon with a net positive charge.

As the positive charges are closer to the balloon than the negatively charged electrons, the force of attraction is greater than the force of repulsion and the can is pulled toward the balloon. The closer the balloon gets, the greater the force, and, therefore, the more the can moves.

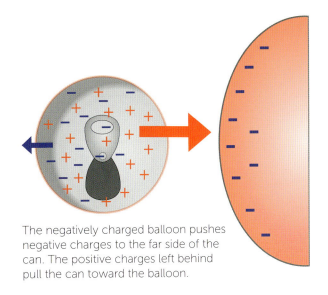

The negatively charged balloon pushes negative charges to the far side of the can. The positive charges left behind pull the can toward the balloon.

HOW does this apply to the real world?

The same science also explains why old-fashioned television screens attract so much dust. The television picture is created by firing a stream of negatively charged electrons at a light-producing coating, known as a phosphor, on the back of the screen. This causes a large negative charge to build up on the glass, similar to the balloon in this experiment.

When airborne dust wafts past and enters the electric field created by the screen, the charges in the particles rearrange themselves like the electrons in the metal drink can. One side of the dust develops a net positive charge that pulls the dust in toward the negatively charged screen. You can demonstrate this effect by tearing up some tissue paper into small pieces and dropping them down the front of a switched-on television set. If they are close enough to the glass as they fall, they will be pulled in and stick.

SOME OTHER THINGS TO TRY
Use a plastic drink bottle instead of a metal can. Does it work? Plastics are generally very poor conductors of electricity, so the charges in the bottle cannot move as easily as they did in the metal can. As a result, the attraction is weaker.

INVISIBLE INK

You often hear about people sending secret messages containing hidden text that can only be revealed by a clever chemical trick. In this experiment, we'll show you how to make a secret ink of your own.

ADULT SUPERVISION!

◆ You will need some paper and a cotton swab, some lemon juice, sugar, water, and a toaster.

◆ Dip the cotton swab in the lemon juice and use it like a pen to write a message on the piece of paper. When it dries, it should be almost invisible.

◆ Switch on the toaster and hold the piece of paper, message side down, in the warm air rising from the toaster. (Do not put the paper in the toaster, and mind your fingers!)

◆ After a short time, you should see your message appear in brown on the paper.

◆ Now try dissolving some sugar in a small amount of water and repeat the experiment using this solution instead of lemon juice. The sugar writing should also appear in brown after a short time above the toaster.

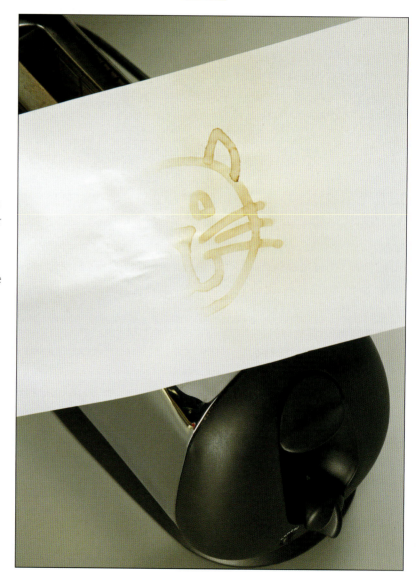

WHY does it work?

It's down to the same food chemistry that makes your dinner taste so tantalizing and turns sugar into toffee. Lemon juice contains a mixture of chemicals including water, sugars, and proteins, and when you write with it, these are absorbed into the surface of the paper. As the juice dries, the water evaporates, but the proteins and sugars are left behind in the shape of the original writing. The protein and sugar crystals are very tiny, and they're also colorless, which is why your dried lemon-juice lettering is very difficult to see.

When you heat the paper in the hot air rising from a toaster, a chemical reaction kicks in between the sugar and the protein, turning the concealed crystals brown, so your writing reappears. This is the Maillard reaction, named after the French chemist Louis Camille Maillard.

What about writing with just a sugar solution, why should that, on its own, turn brown? This is the process of caramelization, a process that is also triggered by heat. At high temperatures, as well as linking up with proteins in the Maillard reaction, sugars can also oxidize and begin to link up with each other, forming sugar polymers. These soak up blue light and that makes them (and your sugary writing) look brown.

HOW does this apply to the real world?

Apart from enabling you to send covert chemical messages, these two reactions are responsible for making cooked food smell and taste delicious. What Maillard discovered in 1912 was that when meats or vegetables are heated to more than 300°F (148.9°C), sugars and proteins naturally present in the food react together to form a family of large brown-colored chemicals called melanoidins. These impart a strong taste and aroma and are the key flavors associated with roasted, fried, or barbecued food. But what about boiled or steamed foods? Water boils at 212°F (100°C). This is not hot enough for the Maillard Reaction to take place, which is why foods cooked by boiling or steaming tend to taste blander in comparison to their grilled counterparts.

And caramelization? This is another taste-enhancing reaction that takes place when food cooks, and it's also how toffee is made. At similar temperatures to those required for the Maillard reaction, sugar molecules first dehydrate, then oxidize and begin to link together forming polymer chains of various lengths. The shorter sugar chains impart tasty caramel flavors to the food, while the longer chains are darker and have a "burnt" taste. They also absorb more blue light than the shorter chains and so tend to look darker. In general, harder, less tooth-friendly toffees contain more of the long polymer chains, which is why they're tougher and darker!

SEEING THE INVISIBLE

Why do stars twinkle and road surfaces seem to shimmer on a sunny day? Why is it that the air above a flame or a hot toaster appears to be twisting and shimmering? In this experiment, we'll find out, and we'll also show you how to turn an invisible gas into something you can see.

◆ For this experiment you will need some bicarbonate of soda (baking soda), some distilled vinegar, a medium sized opaque jug, some white paper, and a clean window with the sun streaming through. If you haven't got a sunny window, then you can use a darkened room and a bright flashlight with the reflector removed instead.

◆ Make sure that all windows and doors are closed to stop any drafts.

◆ Stick the piece of paper to a wall or another vertical surface so that it is illuminated like a screen by the sunlight from the window.

◆ Next, add 3–4 teaspoonfuls of bicarbonate of soda and about ½ inch (1 centimeter) depth of vinegar to the jug. This should froth up, releasing bubbles of gas. When the froth has died down, hold the jug in front of your screen and gently tip it to pour out a thin stream of gas through the sunlight or the flashlight beam. (Remember that the gas will be sitting above the liquid so there is no need to pour out the vinegar.)

◆ As you pour, watch the illuminated part of the screen next to the spout very carefully. Even though nothing appears to be leaving the jug, on the screen you should see wispy light and dark patterns pouring from the spout like an invisible fluid.

WHY does it work?

What you are seeing on the screen is the effect of light changing its speed and being bent when it passes from one substance into another, an effect known as refraction.

When an acid, like vinegar, is added to sodium bicarbonate, it releases carbon dioxide. This is about 50 percent heavier (denser) than plain air, so it builds up in the jug and can be poured out like a liquid.

The heavier carbon dioxide falls downward from the jug as a thin round stream that is shaped a bit like a lens. Since the gas is denser than air, light passing into it slows down very slightly, causing its path to bend or refract.

This has the effect of focusing the light to create brighter patches on the screen, flanked by darker shadows. These are the swirling patterns that you can see.

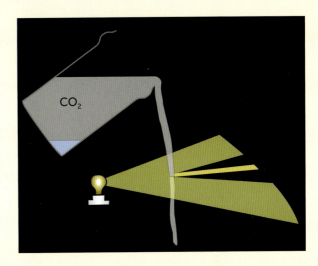

HOW does this apply to the real world?

The same effect is at work on hot days when roads and roofs appear to shimmer in the distance, and it's also why you appear to be able to see "heat" rising above hot toasters, bonfires, and candles. In these situations, hot surfaces heat up the air; this causes it to expand and become less dense. When light passes through this hotter air from the surrounding cooler, denser air, its path is bent, just as it was by the stream of carbon dioxide in this experiment. Since the air is moving, the light is continuously bent in different directions and by different amounts; this distorts the appearances of objects behind it, making them appear to shimmer.

This is also why stars appear to twinkle at night, because the path of their light is warped on its way through the Earth's atmosphere by regions of cooler and warmer air. This blurs the image of the star, so telescopes are often positioned at high altitude or even in space to get around the problem.

THE WORLD'S CHEAPEST CAMERA

How much equipment do you think you would need to produce a full-color moving picture of what is going on outside your window? Well, you can forget all the high-tech stuff, because you can do it just using things you were going to throw away, and at the same time find out how the same science revolutionized 16th-century art.

◆ For this experiment, you will need a large cardboard box (at least 16 x 16 x 20 inches [40 x 40 x 50 centimeters]), a sheet of white paper, some tape, a heavy curtain or other lightproof fabric, and something to make a hole in the cardboard such as a sharp pair of scissors.

◆ Place the box on its side on a table so that the opening points toward you.

◆ Make a hole of ⅛ inch (3–4 millimeters) diameter in the center of one end of the box, and stick the sheet of paper to the inside surface of the box directly opposite the hole.

◆ Now, point the end of the box with the hole at a window or another bright object.

◆ Put your head just outside the opening of the box and cover yourself, and the opening, with the curtain; try to make it as dark as possible inside the box. Look at the paper, what do you see?

The larger the aperture, the brighter and less sharp the image.

◆ On your paper screen, you should see an upside-down and back-to-front picture of what the box is pointing at in full "Technicolor." You have made a pinhole camera that could even take pictures without the aid of a lens (see below). To increase the brightness of the image, try making the hole bigger. The downside is that this will make the image less sharp, so increase the size of the hole by only a small amount at a time.

WHY does it work?

Each part of the surface of an illuminated object reflects rays of light in all directions. Some of these travel toward the box, and a few pass through the pinhole to produce a small colored spot of light on the screen. Rays that enter the pinhole from different objects arrive from different directions, and because light travels in a straight line, they produce spots in different places on the screen.

In this way, many small spots like these build up an inverted image on the screen of what the pinhole is seeing. The image is upside down because light coming from the top of an object above the camera has to travel downward

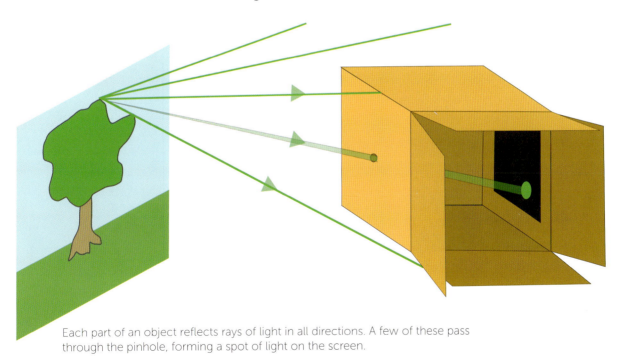

Each part of an object reflects rays of light in all directions. A few of these pass through the pinhole, forming a spot of light on the screen.

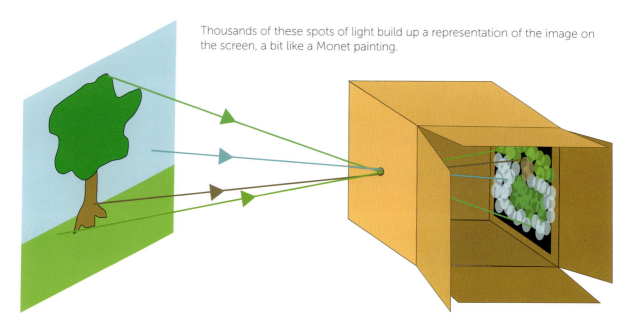

Thousands of these spots of light build up a representation of the image on the screen, a bit like a Monet painting.

to reach the pinhole, and after it has passed through, it continues downward until it hits the screen.

Conversely, light from an object below the camera has to travel upward to meet the pinhole and then continues upward to meet the screen. The same is true for left and right so, viewed from inside the box, the image is projected upside down and back to front.

What happens if you make the hole bigger? When you enlarge the pinhole, more light can pass through; this makes the image brighter. At the same time, the spots of light become correspondingly larger, overlap each other, and make the image less sharp.

HOW does this apply to the real world?

Many art historians believe that pinholes were popular with Renaissance Italian artists who were eager to capture the perfect perspective in their paintings. Some suggest that the famous 18th-century Venetian canal painter Canaletto used a similar technique to produce images of the view he intended to paint. He would then trace lightly around the outline of the image before painting it in, to ensure that his version of the scene was faithful to reality.

IMAGES FROM A MAGNIFYING GLASS

How does a camera focus an image when it takes a picture? And how do our eyes keep our vision sharp? This is down to lenses, and in this experiment we'll show you how they work.

◆ You will need a magnifying glass and a bright window (but not facing the sun) opposite a white wall. If you don't have a suitable wall, stick a piece of white paper to the side of a box and place this makeshift screen on the opposite side of the room facing the window.

◆ Point the magnifying glass at the window and hold it about 6 inches (15 centimeters) from the white wall.

◆ Watch the wall carefully and slowly move the glass backward and forward.

◆ You should see an image of the window and the outside world projected, upside down, onto the wall.

◆ As you alter the position of the lens the image should sharpen. If you continue to move the lens in the same direction the image will become blurred again.

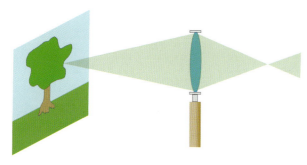

Light from any one point on an object is focused to a point behind the lens, after which it spreads out again.

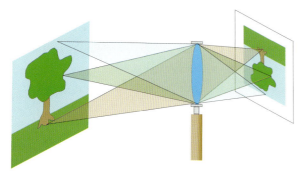

A screen placed at the focal point behind the lens reveals the scene as an inverted image.

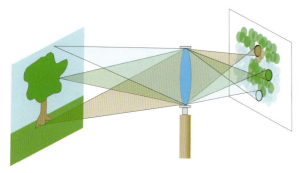

The image on a screen closer to or further from the lens than the focal point is blurred.

WHY does it work?

This experiment is all about optics and the action of lenses, including the ones working right in the front of your eyes! A magnifying glass is a converging lens that is a piece of glass or plastic shaped so that it bends light toward a point, called the focal point, on the other side of the lens. In this way, it concentrates light from a larger area into a smaller area, which is why you see a smaller but brighter version of the view from the window projected onto the wall.

Why does moving the lens backward or forward affect the sharpness of the image? This is because the sharpest image will occur when the focal point of the lens falls at the surface of the screen. If the lens is too close to the screen then the light rays being focused will not meet at a point, and, instead, the image will be blurry. Conversely, if the lens is too far from the screen, the light rays will pass through the focal point and then begin to spread out again, also causing the image to blur.

But why are the images upside down? The image is inverted because light from the top of the window must travel downward to meet the magnifying glass. After focusing, it continues traveling downward on the other side of the lens to form the bottom of the image on the screen. In the same way, light from the bottom of the window travels upward to meet the lens and so ends up at the top of the image.

HOW does this apply to the real world?

This is exactly how both a camera and your eyes work. Inside a camera, the lens gathers light from a wider area and focuses it onto the film or the light-sensitive Charge-coupled Device (CCD) in a digital camera. If you open up an old-fashioned single-lens reflex camera, put a piece of waxed paper where the film should be and hold open the shutter, you can see the image that would be projected onto the film. If you were to replace the piece of waxed paper with some film that changes color when light hits it, then expose it to light for a short period of time before chemically stopping it from changing color any further, you would have a permanent copy of the image—a photograph. This is exactly what a camera does.

Your eyes also work on the same principle, but instead of a piece of film they have a retina that is a layer of light-sensitive nerve cells. These convert the patterns of light that fall onto them into electrical signals that the brain can understand. These signals are used to reassemble a picture of the world, as our eyes see it, inside the brain. But, in the same way that the camera lens turns everything upside down, the images that the brain receives are also upside down. Thankfully, it turns things the right way up before showing them to us!

SOME OTHER THINGS TO TRY

The darker everything else is, the brighter the image will be, so if you can mount the magnifying glass over a hole in a cardboard box the image will look a lot brighter inside the box. Also, if you use a lens to focus an image of the sun onto a piece of paper, the immense amount of light hitting one small spot will cause the paper to heat up, char, and possibly catch fire.

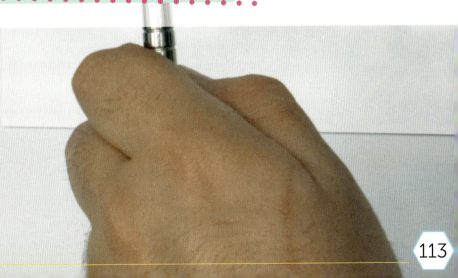

CONFUSE YOUR BALANCE

Most people have found out the hard way what happens if you spend too long on a merry-go-round or have experienced the giddy effects of spinning around in circles. How does the brain's balance system keep you upright, and why don't pirouetting ballet dancers and ice skaters fall over?

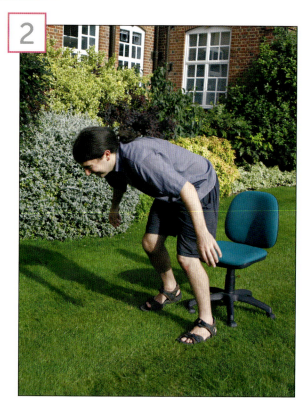

◆ For this experiment, you will need a wide open space with a soft landing, an office chair that can spin in a circle, and a strong assistant.

◆ Sit in the chair facing forward and bend your neck so your head is lying on its side with your ear on your shoulder.

◆ Remain in this position as your assistant spins you in the chair at a constant speed in one direction.

◆ After about 30 seconds, stop the chair and stand up immediately.

◆ Try to walk slowly forward. You might want to make sure your assistant is on hand to catch you, because you will probably fall over, either forward or backward. This is different to when you spin around with your head upright; that tends to make you fall over sideways.

WHY does it work?

It's all down to the brain's vestibular system, a network of fluid-filled canals that can sense movements of the head and trigger compensatory muscle reflexes to keep us on our toes rather than flat on our faces.

In the inner ear on each side of the skull there are three tiny ⅟₂₅-inch (1-millimeter)-wide circular tubes, known as semi-circular canals, that are shaped like miniature ring doughnuts. They are arranged at 90 degrees to each other so that they can detect movements of the head in any direction.

One of the semicircular canals sits horizontally in the same orientation as a halo and signals when the head looks left and right. The second sits vertically, pointing from the front to the back of the head and detects nodding movements, while the third, that also sits vertically, points from one ear toward the other and signals when the head leans over toward the shoulder.

But how do they do this? The canals are filled with a watery liquid and whenever the head moves the liquid is initially "left behind" inside the canal and does not move with the rest of the tube. This causes it to press on a cluster of tiny hairs that extend into the fluid from the surface of the canal. Movements of these hairs trigger the production of nerve impulses, and by comparing the combined signals from all three semi-circular canals on each side the brain can work out in which direction the head is moving and how quickly.

When you sit in the chair and spin in the same direction for a long period of time, the liquid in the semicircular canals also begins to turn. And because your head is on your shoulder, the vertical canal that normally detects nodding movements is also lying on its side, so the liquid inside begins to spin in the same direction as the chair. This means that when the chair stops spinning and you stand up, the fluid in this canal continues to swirl around, fooling the brain into thinking that you are falling forward or backward (depending upon what direction you were spinning).

In an attempt to prevent you from toppling over, the nervous system triggers muscle reflexes to compensate for the perceived movements, but as you aren't actually falling over, this is the wrong response, and it has the reverse effect, which is why your assistant had to catch you!

HOW does this apply to the real world?

The brain relies on the vestibular system not only to keep us upright but also to help us to see clearly. To demonstrate this, try holding up a finger in front of you and shaking it quickly while trying to follow it with your eyes. It's impossible to see objects clearly if the eye cannot fix on them for long enough because they are moving.

But now hold your finger steady, fix on it with your eyes, and shake your head rapidly from side to side or up and down. Your finger remains visible all the time, and if you watch someone else doing this, you will see that their eyes are continuously compensating for their head movements so that they can remain fixed on the finger.

This is known as a vestibular-ocular reflex; your semicircular canals trigger eye movements in the opposite direction to head movements so that you can still see while you are running, driving, or playing sports.

That's all very well, but how do pirouetting ballet dancers and ice skaters avoid becoming dizzy? They use a technique called "spotting." If you watch them carefully you will see that they fix their gaze on a conspicuous object and follow it with their eyes until the last possible moment at which point they whip their head around quickly and fixate upon the object again. This minimizes the time the head spends turning, so it reduces the likelihood of dance-destroying giddiness and disorientation!

SOME OTHER THINGS TO TRY

If you have a clear, safe, grassy slope that's not too steep, try doing forward rolls down it. When you get to the bottom you will feel as if you're still rolling forwards. Also, try repeating the spinning chair experiment but with your head on the other shoulder—does it make a difference? Now spin the other way—what happens?

DIY BUTTER

Ask most people where butter comes from, and they'll say "the supermarket." But how is butter really made? In this experiment, we'll show you and also explain how the same science can help with DIY and decorating.

- ◆ You will need some heavy cream, a bowl, a whisk, and some paper towels.

- ◆ Pour the cream into the bowl and whisk it briskly. It should soon become thicker and stiffer and perfect for spreading.

- ◆ Don't stop at this point: keep whipping. Before long, the cream will begin to thin again and turn yellow.

- ◆ Scoop it out onto some paper towels and squeeze out the excess water. The result inside the towel should look and taste like butter ... because that's what it is!

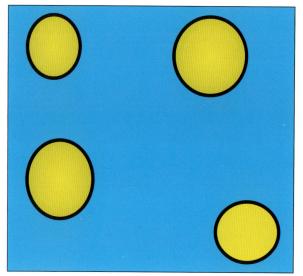

Milk comprises globules of fat (yellow) suspended in water mixed with proteins (caseinogen).

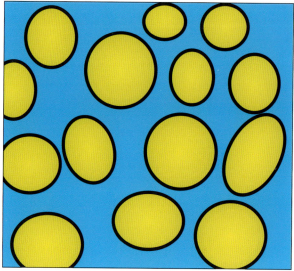

In cream, the proportion of fat is higher.

WHY does it work?

It's not just paint that comes as an emulsion, cream does too! Milk is about 5–10 percent fat, and because fat is less dense than water, if milk is left to stand for any period of time, the fat rises to the top, producing cream. This is "skimmed off" to leave lower-fat milk, and the cream, that is about 40 percent fat, is sold separately.

Under a microscope, cream consists of millions of tiny globules of fat floating in water; this is known as an emulsion. The fat globules are covered by a protein called caseinogen which makes them negatively charged; this keeps the globules apart as like charges repel each other.

When you whip the cream, the globules of fat begin to link up and partially merge together to form clusters resembling tangled strings of beads surrounded by water. These clusters of fat trap air bubbles, which is not only why whipped cream becomes fluffy and "light" but also why it increases in volume. The cream also becomes much stiffer because the larger clusters of fat globules find it more difficult to flow past each other.

Whipping the cream beyond this point causes the fat globules to merge together completely and form oily solids that separate from the water. Straining off this water, known as buttermilk, leaves behind the butter that consists of mostly fat with a small amount of water and milk protein locked away inside.

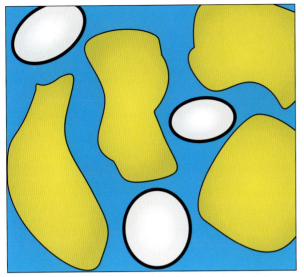

Whipping the cream causes the fat globules to join together and also introduces air (white).

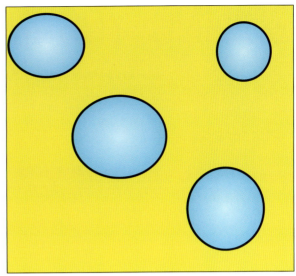

Butter is mainly fat with small amounts of water trapped inside.

HOW does this apply to the real world?

This is exactly how the butter that you spread on your toast is made, although thankfully a machine does the whipping rather than a person! This wasn't always the case, however, and before machines, butter was made by milkmaids who stirred cream in wooden churns. Today, we also have refrigerators to keep our dairy products fresh, but historically people made butter because the oily surface kept out oxygen and slowed down the growth of microbes.

The same science also comes in handy in DIY and decorating. Emulsion paints consist of tiny globules of paint dissolved in oil and suspended in water. This means that the paint behaves as though it were water-soluble and is easy to

apply and wash off. But, after it's been painted onto a wall, the water evaporates leaving the oil-based, water-resistant paint sticking firmly to the surface. Just don't whip your emulsion paint before you use it!

SOME OTHER THINGS TO TRY

Use the same science to make some salad dressing: mixing oil and vinegar quickly results in two separate layers, but if some mustard is included and the three are whipped together, it can stabilize the mixture just as the caseinogen does in milk.

BOILING YOGURT CONTAINERS

Plastics are also known as polymers, but how do they work and what happens when you give them some energy? In this experiment we'll show you, and explain why yogurt containers are best served cold.

ADULT SUPERVISION!

◆ You will need some empty yogurt containers, a saucepan filled with water, some salt, and a large spoon.

◆ Heat the pan of water on your stove.

◆ As it warms up, add several tablespoonfuls of salt to the water. This will raise the temperature at which it boils to above 212°F (100°C).

◆ Once the water is boiling, carefully add your yogurt container and leave it in the pan for about 3 to 4 minutes.

◆ Then remove it with the spoon and place it on the worktop to cool down. The container should have shrunk into itself to produce a thick, flat piece of plastic much stiffer than the original container.

WHY does it work?

The yogurt container is made of a thermoplastic that flows and deforms when it's warm but becomes stiff and brittle when it gets cold.

Plastics are polymers consisting of millions of atoms linked together like long strings of beads.

Initially these polymer chains arrange themselves into compact wiggly tangles that need to be unwound and stretched out in order to produce molded shapes like yogurt containers.

To do this, the plastic is heated to 230°F (110°C). As the temperature rises, the polymer chains begin to shake and writhe around; this makes it much easier for them to slip past one another, and the plastic becomes floppy and deformable. At this point, the plastic can be stretched into a desired shape; this forces the polymer chains to unwind and become elongated and straight.

Then, to ensure that the plastic remains in this new configuration, it is quickly cooled down. This locks the polymer chains into their new positions. But if the plastic is later reheated to a similar temperature, the polymer chains can wriggle back to their previous compact, tangled shapes. This is why the yogurt container shrinks in on itself and produces a much thicker, flatter, stiffer piece of plastic when it is immersed in the boiling water.

Plastics are polymers made from millions of atoms linked together to form long chains.

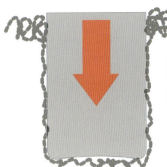

A "former" is pushed into the warm plastic to stretch out the polymer chains.

When cold, the molded plastic retains its new shape.

When heated, the polymer chains begin to shake and writhe around...

... eventually returning to their previous compact, tangled shape.

HOW does this apply to the real world?

Exactly this technique is used to produce many of the plastic objects we see around us. Sheets of thermoplastic are heated so that they become soft and deformable, and then an object of the required shape, known as a former, is pressed into them. The air is then drawn out from around the former, pulling the plastic tightly over the surface so that it takes on the same shape. This is known as vacuum forming. Once the plastic is the correct shape, it is cooled down to fix the stretched-out polymer chains into position, and—hey presto!—it's a yogurt container, or even a car fender.

SOME OTHER THINGS TO TRY

Try the experiment with other plastics, such as candy wrappers or chip bags/crisp packets to find out if they behave the same way. Also, try drawing something on the yogurt container in permanent marker before you heat it, then watch as the drawing changes shape.

FLYING TUBES

When David Beckham spectacularly plants the ball in the net from a corner, at first it often looks like he's going to miss, but then the ball suddenly veers into the goal. Although he might not realize it, he's actually performing a feat of physics that makes baseballs curve and will even make a kitchen paper roll take to the sky.

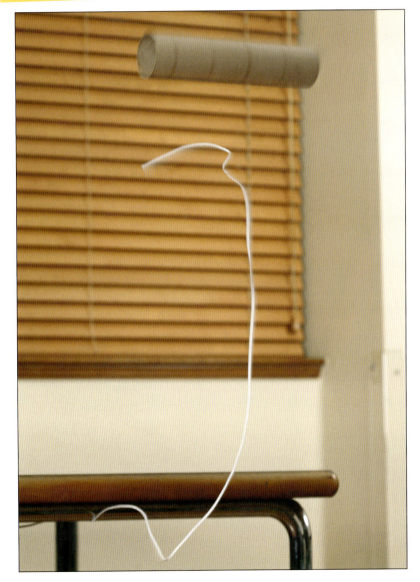

◆ For this experiment you will need a cardboard tube from the center of a kitchen paper towel roll, 2–3 yards (2–3 meters) of narrow elastic, a table, some tape and a large room.

◆ Attach the elastic to the center of one end of your table with the tape so that the free, unstretched elastic is about two-thirds of the length of the table. This is your runway, so point it away from anything breakable!

◆ Stretch the elastic to the end of the table and roll six to seven turns onto the center of your tube (roll the elastic over itself on the first couple of turns to trap the end against the cardboard).

◆ Wind it as though it were a carpet you were going to roll out; in other words, the elastic needs to leave the cardboard on the underside of the tube.

◆ Now hold the tube at the center and let go! The roll should take off and curve upward as it travels across the room.

WHY does it work?

- The roll takes to the sky thanks to the Magnus Effect, named after the German scientist Heinrich Gustav Magnus who studied the aerodynamics of spinning cylinders and spheres in the 1850s. The same science also explains David Beckham's uncanny corners as well as power-serves at the US Open.

- When the tube takes flight, passing air sticks to the curved surfaces and then detaches behind the tube. This is known as the Coanda Effect, and normally the way air sticks to the top and bottom of an object is identical. However, as the elastic unrolls, it causes the tube to spin, with the bottom of the tube turning toward the direction in which the tube is traveling and the top of the tube turning away.

- This means that air finds it easier to stick to the top of the tube, that is moving in the same direction as the oncoming air, than to stick to the bottom, which is moving against it.

- As a result, the air flow tends to detach from the tube earlier on the bottom than on the top, meaning that, overall, the air is being pushed down around the tube.

- Isaac Newton's Third Law tells us that for every action there is an equal and opposite reaction. If the air is being pushed down, the tube must be pushed upward, and so it flies upward. This is why it takes off so spectacularly.

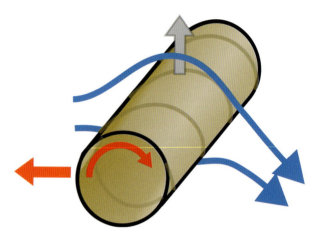

Air sticks better to the upper surface of the spinning tube, because it is turning in the same direction as the passing air, creating lift.

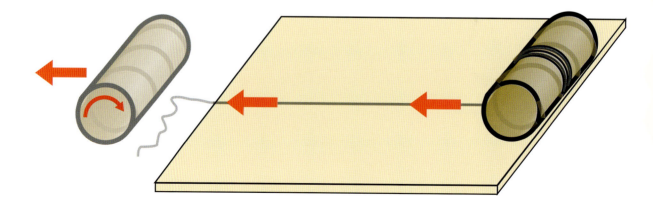

HOW does this apply to the real world?

Just as the cardboard tube in this experiment lifted up as it flew, the Magnus Effect can also cause spinning balls to change direction, often in a way that's difficult to predict. For example, kicking a soccer ball slightly to the side of its center point will make the ball turn as it flies through the air. This will cause the passing air to stick for longer to the side of the ball that is turning in the same direction as the passing air. The result will be a force pushing the ball toward the direction of spin, causing it to follow a curved trajectory.

This doesn't happen throughout the ball's flight. Instead, it becomes more pronounced as the ball slows down, because at low speeds it is easier for air to stick to the ball's surface, so the Magnus Effect is stronger. This is why Beckham's corner kicks are so lethal. He taps the ball at just the right speed so that, as it nears the goal, it has slowed down enough for the Magnus Effect to kick in and curve the ball into the net.

And at the US Open? Tennis players use the Magnus Effect to help them to deliver high-speed serves yet keep the ball in the court. They apply top spin to the ball so it rotates in the opposite direction to your tube in this experiment. This pushes the ball downward as it slows down, so it drops just inside the baseline, fast and deadly!

Why are these trajectories so hard to predict? One theory is that because spinning objects don't generally occur in nature, the human visual system has evolved to account only for the effects of gravity, leaving us powerless against the Magnus Effect on the sportsfield!

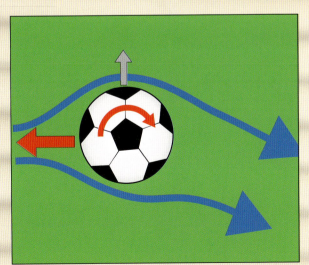

At high speeds (Beckham's initial kick), the air cannot easily stick to the surface of the spinning ball so it flies in a fairly straight line.

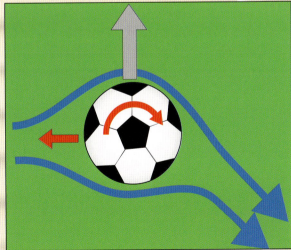

As the ball slows down (as it nears the goal), air sticks to the surface much better, bending its flight into the net.

MAKE YOUR OWN GULF STREAM

What drives the warm ocean currents that bring nice weather north to Europe, and how do bartenders mix multi-layered cocktails?

◆ For this experiment, you will need a smooth-sided glass bowl, some water, and two brightly colored drink concentrates—one very syrupy, such as blackcurrant, and one only quite sugary, such as lime cordial.

◆ Half-fill the bowl with water.

◆ Place it on a flat surface, and then leave it to stand for 10 minutes so that the water settles down and stops moving. If the water is not still, then the experiment will not work.

◆ Now, take the runnier of your two drink concentrates and very gently pour about one quarter of a cupful down the side of the glass bowl into the water. The liquid should hug the glass and run beneath the water to form a visible layer at the bottom of the bowl. Notice how the water swirls slightly as the concentrate flows underneath it.

◆ Wait a minute or two for the liquids to settle down again, and then repeat the process with the second, thicker concentrate. It should flow beneath the first drink concentrate and form a second layer at the bottom of the bowl.

◆ If you now look through the side of the bowl, you should able to see three liquid layers stacked above each other, just like a fancy cocktail.

◆ Finally, to prove that the liquids can all mix together and it's not a trick, take a spoon and stir the bowl: the layers will vanish and you will be left with a single solution.

WHY does it work?

This experiment is all about density and why liquids of different densities don't easily mix. It's the same process that drives the ocean currents that send cold water from the Arctic down to the Equator and bring warmer water back.

Drink concentrates contain large amounts of sugar dissolved into the fruit juice. When the two mix, the sugar molecules fit into spaces between the water molecules in the juice. Although adding the sugar makes the juice much heavier, it doesn't take up much more space. This means that a given volume of the mixture weighs more than the same quantity of water. In other words, the concentrate is denser than water, so it sinks to the bottom of the bowl.

If you add a different concentrate that has even more sugar dissolved into it, this will be denser still and will sink right to the bottom of the bowl below both the water and the first concentrate layer. If you were to leave this combination of liquids perfectly still and didn't stir it or heat it, it would take thousands of years for molecular vibrations to mix the liquid layers together.

HOW does this apply to the real world?

Britain and northern Europe are much warmer than some parts of Canada, despite being as far north. This is because they are warmed by an ocean current called the Gulf Stream, that comes from the Gulf of Mexico and brings with it heat equivalent to the output of one million nuclear power stations.

By the time it reaches the Arctic, the water has cooled down and begins to freeze. Since ice is mainly pure water, the remaining seawater becomes much saltier; this increases its density and causes it to sink to the ocean bottom, just like the drink concentrate in this experiment. The cold salty water then flows back south across the floor of the Atlantic. This sinking water draws fresh water north again: without it, Brits would all be reaching for a fourth sweater!

And the fancy cocktails you see in expensive bars? These are made the same way. Bar staff use mixers of different concentrations to achieve colorful layer combinations.

MUSIC FROM A WINE GLASS

Almost anything can be used to make music, including a wine glass. In this experiment we'll show you how the humble glass can be used to demonstrate the principles of resonance and why some music makes your windows rattle.

◆ You will need some wine glasses, some water, and your finger.

◆ Lightly flick the rim of each glass in turn.

◆ Pick out the one that "rings" the longest.

◆ Next, place the glass on a flat surface, moisten your finger, and rub it gently around the rim of the glass, taking care not to press too hard. It may help to support the base of the glass with your free hand. With practice, you will be able to make the glass produce an eerie note rather like a bow being drawn slowly across the strings of a violin.

◆ Now, add some water to the glass and "play" it again. You should notice that the pitch of the note drops.

◆ Add more water, and it should drop further.

HOW does this apply to the real world?

This experiment is all about resonance and the way objects vibrate. This principle also spawned an unusual musical instrument in the 1700s that allegedly made people go mad.

Many objects have one or more frequencies, known as resonant frequencies, at which they vibrate best. This is why pushing a swing at just the right moment (in other words, with the right frequency) makes it swing higher, and why older cars often develop a rattle at certain engine speeds.

A wine glass is the same, and when you put energy into it by running a finger around the rim, the material vibrates at the resonant frequency of the glass, which is the note you hear. This occurs because your finger does not make a smooth movement over the glass surface. Instead, it moves in a series of short jerks, sliding a short distance and then stopping momentarily, and this is the source of the vibrations that cause the glass to resonate.

Cheap glasses tend not to work as well as their expensive crystal counterparts because they often contain minute flaws that rub against each other and damp down the vibrations. This is like dragging your feet across the ground when you are on a swing: it wastes energy and slows you down.

When it works correctly, the glass makes a sound because the vibrations are transmitted to the air around the vessel, creating sound waves. Why does the note change when water is added? This is because the water makes the glass heavier and that slows down the rate at which it can vibrate so that the sound waves it produces have a lower frequency (pitch), in the same way that heavier guitar strings produce lower notes than lighter strings.

HOW does this apply to the real world?

In the 1700s, the scientist Benjamin Franklin invented a musical instrument based on resonating glasses. It was called a glass harmonica, or armonica for short, and it used a foot pedal to rotate multiple glasses, all with different resonant frequencies so they could produce different notes. The instrument was played by applying damp fingers to the rims of the turning glasses to create sounds like the ones you made in this experiment. The rims of the glasses were color-coded so that the player could tell the notes apart, and by playing multiple glasses at once it was possible to create chords. Mozart and Beethoven even composed dedicated pieces of music for the instrument.

The story goes that people who played or listened to the glass harmonica were at risk of going insane. While this is almost certainly not true, it is possible that lead from the colored paints used to identify the different glass "notes" could have rubbed onto the fingers of the player. As the fingers needed to be constantly licked to keep them moist, this may have resulted in a reasonable intake of lead and that is known to be bad for the brain!

Although the glass harmonica may have fallen from fashion, almost everyone today has a stereo system at home, and most people can think of at least one piece of music that, if played loudly enough, will make objects in the room vibrate noisily. This is an example of resonance. Certain pieces of music are dominated by notes at a specific frequency, and if this coincides with the resonant frequency of some of the things in your living room, you will hear them shaking!

SOME OTHER THINGS TO TRY

Put some water in the bottom of the glass, and rub the rim to make it resonate. Stop rubbing, but while the glass is still ringing, pour the water out. You should hear the pitch move downward. This is because the water adds mass to the part of the glass that is moving the most, slowing down the vibrations.

THE SCIENCE OF TIDAL WAVES

Fans of George Clooney will recall him and his crew being swallowed up by a giant wave in the movie *Perfect Storm*, but can swells like this really happen? The answer is yes, and this experiment will show you how.

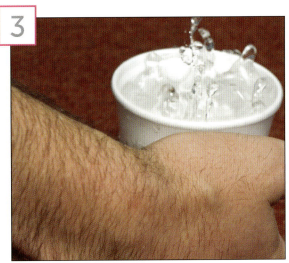

◆ You will need a styrofoam cup, some water, some coins to use as ballast, and a carpeted surface.

◆ Place the coins at the bottom of the styrofoam cup; they are there to add weight and make the cup more stable.

◆ Now fill it with water to a depth about ½ inch (1 centimeter) from the rim and place the cup on the carpet.

◆ Using your index finger, push the cup along by its base at a steady speed and watch the surface of the water as the cup moves.

◆ Within a short time, the water surface will become turbulent and should begin to generate waves that spit water out of the cup.

WHY does it work?

This is the science of superposition: how waves can add together to cancel each other out, or produce even bigger waves. When you push the cup along the carpeted surface, it will repeatedly stick and then slip. This sets up vibrations in the sides of the cup. These are transmitted to the water and create a series of circular waves that travel across the surface and then reflect off the opposite sides of the cup.

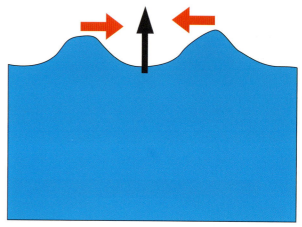

As waves converge on a point, sometimes they add together to create a much bigger wave that can splash from the cup.

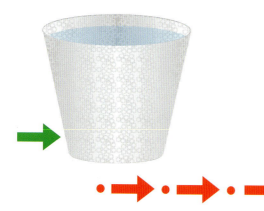

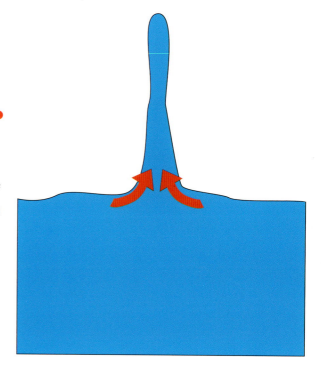

Wherever the waves meet with other waves coming from different directions they merge or "superpose." When this happens, if the trough of one wave coincides with the peak of the second wave, the two cancel each other out. But if two wave peaks coincide, they add their heights together to produce a bigger wave.

As the cup moves across the carpet, a pattern of flat spots and rough spots emerges on the surface of the water as the waves superpose. If enough wave peaks converge at the same place, they can produce a wave powerful enough to spray water out of the cup.

HOW does this apply to the real world?

The same phenomenon might explain why ships go missing at sea and how ultrasonic smoke machines work. Scientists now think that wave superposition might explain the occurrence of freak massive waves over 100 feet (30 meters) tall that have been known to sink ships and damage oil rigs. Previously dismissed as exaggerated stories by sailors, in recent years, waves on this scale have been recorded first hand by research ships in the major oceans. For instance, in February 2000, the Royal Research Ship *Discovery* that was studying ocean storms was battered for 12 hours off the coast of Scotland by the tallest waves ever measured; some were up to 95 feet (29 meters) tall.

Exactly how these giants form is still a mystery, but the convergence of ocean currents bringing waves from several directions might be one reason. It has also been suggested that larger, faster moving waves can outrun and "collect" several smaller waves to make a monster. Whatever the cause, conditions like these, that were previously written off as science fiction, could well explain the disappearance of a number of boats that have vanished without trace in the past.

And the smoke machine? You have probably seen displays at garden centers and aquatic shops that create a layer of white mist that sits above the surface of the exhibit. This is the artificial equivalent of a cloud, and it's made using a similar principle to the cup and carpet in this experiment. In this case, an ultrasound probe is used to vibrate a small patch of water 1.5 million times per second. This causes the water to form waves that spit out tiny droplets of water that hover in the air as a mist. Eventually they rejoin with other droplets and then fall back into the water, but they look good while they last.

SOME OTHER THINGS TO TRY

Add some dishwashing liquid to the cup. You should be able to produce some even bigger waves, because the detergent cuts down the surface tension of the water.

THE SCIENCE OF POOL

The science of spin plays a key role in ball sports, and in this experiment we'll show you how and also explain why a pool referee spends so long cleaning the balls.

- You will need a bouncy ball, some cooking oil or petroleum jelly, and a kitchen worktop with a flat surface meeting a solid vertical wall at the back. These surfaces should be smooth but not slippery; a tiled surface is ideal.

- Begin by rolling the ball across the worktop toward the wall so that it hits the vertical surface and rolls back toward you.

- Do this several times and take note of how the ball travels and how it turns as it rolls. (Using a ball with stripes or some dots painted on it can help to make the ball's movements easier to follow.)

- Next, rub some of the oil or petroleum jelly onto the vertical surface at the height where the ball was first hitting the wall.

- Now roll the ball at the lubricated patch of wall and compare how it behaves.

- It should roll back much more slowly from the lubricated surface than the dry wall, and if you watch it turn, you may notice something else happening.

WHY does it work?

It's all to do with the direction of spin, and the same science is at play in billiards and pool. When a ball is rolling, it has two types of momentum: forward motion and rotation (spin). To change direction, such as bouncing back from a wall, both of these have to be altered. When the ball hits the dry (unlubricated) vertical wall, it bounces back just as you would expect, but it also stops spinning.

This occurs because, when it reaches the vertical wall, the turning ball rolls up the surface a short distance and this soaks up most of its rotation. It then drops back to the surface and at the same time bounces back, picking up a new direction of spin as it rolls toward you.

When the wall is lubricated with oil, the slippery surface prevents the ball from rolling up the wall and reducing its spin. As a result, it bounces back but is still spinning as if it were rolling away from you.

This means that friction will continue to slow it down until it begins to roll in the correct direction again, and by this time it is moving much more slowly. If your ball has stripes or lines, you should be able to see the direction of spin switch round.

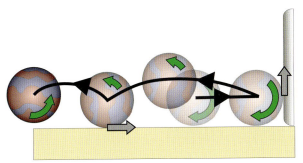

A ball hitting an unlubricated surface rolls up the wall a short way, soaking up its spin so it rolls back much more quickly.

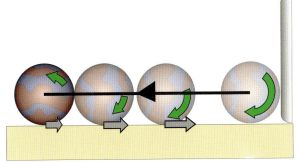

If the wall is then lubricated, it stops the ball rolling upward so it bounces back still spinning the wrong way. This slows it down.

HOW does this apply to the real world?

The physics in this experiment is award-winningly crucial to pool players. When two balls strike each other they should instantly separate after the collision. However, dirt on their surfaces can cause the first ball to roll up the side of the second as it hits. This is known as a bad contact and can cause the second ball to spin in the wrong direction and slow down very quickly. To prevent this from happening, the referee regularly cleans the balls to remove any traces of dirt or chalk.

THE SCIENCE OF BELLS AND COFFEE CUPS

When you stir your morning tea, have you ever noticed how the spoon seems to make a different sound depending upon where it strikes the side of the mug? If so, your ears aren't deceiving you, it's true, and this experiment will reveal why.

◆ For this experiment you will need only an empty china mug and a spoon.

◆ Place the mug on a flat surface.

◆ Tap the rim in different places with the end of the spoon.

◆ Listen to the sound it makes.

◆ If you work your way around the edge of the mug, you will see that when you hit it at 90 or 180 degrees to the handle on either side (1), it produces a lower note than when you hit it at 45 or 135 degrees to the handle (2).

WHY does it work?

This is all about sound and resonance, and the same science explains why the bass strings of a guitar are much thicker than the strings that produce high notes and why some bells seem to pulse as they ring. When you hit the edge of the mug with the spoon, it makes a sound because energy from the strike causes the porcelain to vibrate. These vibrations are transferred to the air where they form sound waves that your ears can pick up.

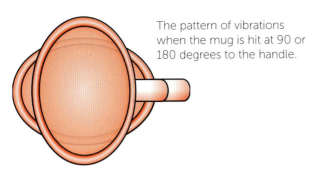

The pattern of vibrations when the mug is hit at 90 or 180 degrees to the handle.

When struck at 45 or 135 degrees to the handle, the mug produces a higher note because it vibrates faster.

You cannot see the mug vibrating because it is happening too quickly, but if you could watch it in slow motion, you would see the circular rim of the mug become squashed into a slight oval shape when the spoon hit. It would then rebound and form an oval at 90 degrees to the first one. This process continues until all of the energy from the spoon strike turns into sound waves and heat. When you strike the mug perpendicular to, or in line with the handle, the handle has to move as the body of the mug compresses and stretches. This makes the moving parts of the mug heavier, so they vibrate more slowly and produce a lower frequency sound.

When the mug is hit at 45 or 135 degrees to the handle, the compression and stretching does not involve moving the handle, so the moving parts weigh less, vibrate more rapidly, and produce a higher pitched sound.

HOW does this apply to the real world?

The same piece of physics explains the basis of bass guitars. They have much thicker, heavier strings, that vibrate more slowly to produce lower frequency (lower pitch) sounds than an instrument with thinner strings.

Another place where you might see the science of this experiment in action is in a bell tower. Large bells are very rarely perfectly circular, so they behave like the mug and handle in this experiment. The bell has to be positioned so that the clapper strikes it in just the right place to obtain the correct note. If not, it can produce two notes—one higher and one lower—known as a doublet, every time it is hit.

You can demonstrate this with your coffee mug by tapping around the rim. You will find a point halfway between 45 and 90 degrees from the handle where both notes are audible.

This is also why some electric bells, when rung continuously, can cause an unpleasant pulsing effect where the sound becomes louder and softer. This occurs because the frequencies of the two notes from the bell are different, so periodically the peaks of their two sound waves will coincide with each other, producing a louder sound.

JAM JARS AND FLYWHEELS

What's heavier, a pound of feathers or a pound of lead? It's a trick question, of course, but there are examples in science when two things with the same weight can behave quite differently, like a jar of jam and a jar of water. In this experiment, we'll use them to show you how a flywheel works and what powers an ice skater's pirouette.

- For this experiment, you will need a table about 6 feet (2 meters) long and three identical jars with their tops, two empty and one containing peanut butter or jam.

- Prop up one end of the table so that it forms a sloping ramp with one end about 1½ inches (4 centimeters) higher than the other.

- Put some heavy books across the bottom end to act as a stop so that things cannot roll onto the floor.

- Fill one of your jars with water and leave the other empty.

- Tighten the tops on both.

- Now, place the three jars on their sides in a line at the top of the ramp and race them down the slope to see which rolls to the bottom first. If you find it difficult to release the jars at the same time, it may help to use a piece of wood as a starting gate. Place it below the jars so that it stops them rolling and then, holding it at both ends, quickly pull it out of the way down the slope.

- Run a few trials to confirm your results, swapping the starting positions of the jars.

- You should see an interesting pattern emerge: the water-filled jar should always win, followed by the peanut butter or jam. The empty jar will limp in last.

WHY does it work?

You might expect them all to reach the bottom at the bottom at the same time, like Galileo's two stones, one heavy and one light, that he dropped from the Leaning Tower of Pisa. But your jars are behaving differently, and they are showing you the science of spin, and how flywheels store energy to keep an engine running smoothly.

Objects are accelerated downward by gravity, and a heavier object is pulled harder than a lighter one. Because a heavier object, by definition, has more mass, it requires a greater force to accelerate it, and so the two effects cancel each other out. This is why a light and a heavy stone, dropped side by side from a tall building, will hit the ground together.

However, rather than just falling, the jars in this experiment are also moving in an additional way:

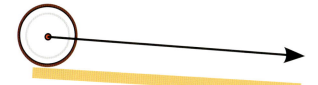

Mass at the center of the jar moves in a straight line as the jar rolls along.

Mass at the edge of the jar has to be accelerated along a curved path, as well as moving down the slope.

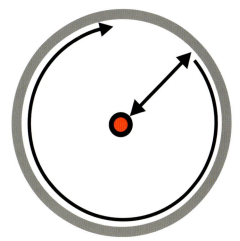

In an empty jar, most of the mass is around the outside.

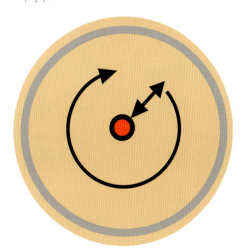

In a peanut-butter-filled jar, more mass is located close to the center.

they are rolling and that means that at least some of their mass also has to be accelerated into a circular motion. The greater the proportion of their mass that has to be accelerated in this way, the longer they take to speed up.

The empty jar loses the race because most of its mass is in the glass around the outside of the jar, and, therefore, it all has to be accelerated into a rolling motion. The jar filled with peanut butter comes second because it weighs more than the empty jar, so gravity pulls it harder, but not all of its mass is round the edges. The peanut butter toward the center of the jar doesn't have to roll very far but nevertheless helps to pull the jar downhill.

The water-filled jar wins because the water adds mass that accelerates the jar downhill, but the water doesn't rotate inside the jar: the glass turns around it, and the water stays still. As a result, only a small proportion of the jar's weight has to be accelerated into a rolling motion, so the water jar makes it to the bottom first.

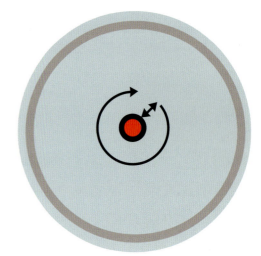

In a water-filled jar, the water doesn't move, so it behaves as though all of its mass is at the center.

HOW does this apply to the real world?

This is how a flywheel works to store energy to keep an engine running smoothly between the intermittent power-strokes from the pistons, and how ice dancers deliver dazzling pirouettes.

Flywheels have the majority of their mass around the rim, and as an engine speeds up, the flywheel is accelerated, storing energy. Between one piston stroke and the next, the flywheel feeds some of this energy back into the engine to keep it turning smoothly, especially at low speeds.

And ice skaters? You may have noticed that an ice skater can speed up a pirouette if they pull their limbs in toward their torso. This is the equivalent of turning the peanut-butter-filled jar into the water-filled jar. With their arms outstretched, they are moving some of their mass in a very large circle, but with their arms tucked in close to their body, most of their mass is moving in a much smaller circle, so they turn more quickly.

SOME OTHER THINGS TO TRY
Compare dropping a yo-yo (while holding the string) alongside a second free-falling object. The yo-yo has to develop a circular motion like the jars in this experiment, so it accelerates more slowly than the free-falling objects.

WATERPROOF HANKY

When you want to go out in the wet and stay dry, you put on a raincoat. But unless it's "breathable," before long the build-up of sweat inside the jacket will render you almost as wet as if you hadn't worn anything! How do these waterproof-but-breathable fabrics work? In this experiment we'll show you.

◆ You will need a cotton handkerchief, a glass, and some water.

◆ First, stretch the handkerchief over your mouth and breathe lightly in and out. You should find that air from your mouth can easily flow through the open weave of the fabric.

◆ Now fill the glass to the brim with water and stretch the handkerchief over the top.

◆ Hold it tight against the sides and then quickly turn the glass over and keep it upside down.

◆ If you're feeling brave you can even try turning it upside down above someone's head, but don't let go! If you've got it right, the water should remain in the glass, behind the handkerchief.

WHY does it work?

This experiment relies on the science of surface tension and also explains how fabrics full of holes can be both breathable and waterproof at the same time. When you invert a liquid-filled glass without a handkerchief across the top, the liquid falls out and is replaced by a succession of large air bubbles that rise to the top of the glass to fill the space vacated by the falling water.

But if air cannot enter, a vacuum forms above the liquid, preventing it from falling out.

This is exactly what the handkerchief does. Although it contains thousands of tiny holes between the fibers, these become covered with a layer of water. For air to get in and allow the water to fall out, it would have to form tiny bubbles through the mesh of the handkerchief.

Just like a balloon, that is hardest to blow up when it is at its smallest, the tinier a bubble is, the harder the water's surface tension tries to make it collapse. The bubbles that would allow air to pass through the handkerchief are so small that the air cannot push hard enough to form them against the collapsing force, so the water stays in the glass behind the hanky.

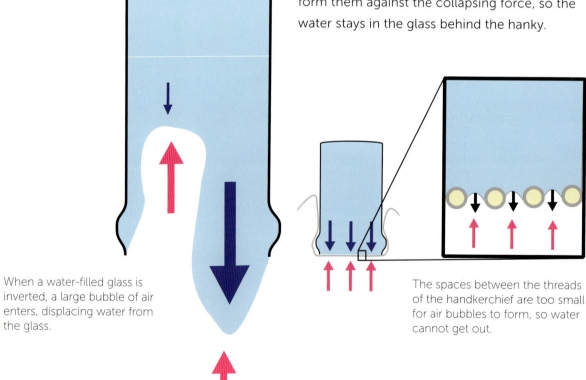

When a water-filled glass is inverted, a large bubble of air enters, displacing water from the glass.

The spaces between the threads of the handkerchief are too small for air bubbles to form, so water cannot get out.

HOW does this apply to the real world?

The same science can be used to explain the basis of breathable fabrics, like Gore-Tex™. This consists of a thin membrane sandwiched between a water-repellent outer layer and a warm inner layer. The membrane is peppered with billions of tiny pores that are 20,000 times smaller than a raindrop but thousands of times larger than a water molecule.

For raindrops to move from the surface of the coat through to the inside and make you wet, they would have to break up into tiny droplets and, just like the air bubbles crossing the handkerchief, surface tension prevents them from forming. However, individual water molecules evaporating from your sweat can easily pass through the pores to the outside, so your sweat can escape, but the rain is kept out.

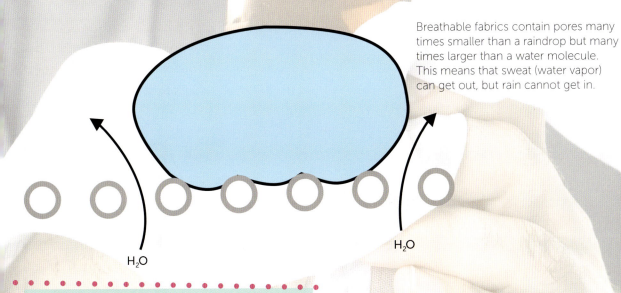

Breathable fabrics contain pores many times smaller than a raindrop but many times larger than a water molecule. This means that sweat (water vapor) can get out, but rain cannot get in.

H_2O

H_2O

SOME OTHER THINGS TO TRY

Surfactants like soaps and detergents reduce surface tension by forming a surface layer that water molecules can bond with, so try adding detergent or using a less tight weave cloth to see when your handkerchief stops being waterproof.

143

Dr. Chris Smith is based at Cambridge University where he is a lecturer and a medical consultant. He is also the founder and presenter of the multi-award-winning *Naked Scientists* radio show and podcast, which is one of the world's most downloaded science

Photo credit: Jason Hudson

programmes. Chris also appears live every week to talk science and answer listener questions on radio networks internationally including on the BBC, the ABC in Australia, Radio New Zealand and Primedia's Talk Radio 702 in South Africa. Chris lives near Cambridge, UK, with his family.

Dave Ansell is a physicist by training but did so much science outreach in his spare time that he never finished his PhD. He has managed to get paid to do his hobby creating science museum exhibits for the Cambridge Science Centre and recording

Photo credit: Cambridge Science Centre

hundreds of *Kitchen Science* experiments with the Naked Scientists, the best from which appear in this book.

Published 2016— IMM Lifestyle Books
www.IMMLifestyleBooks.com

IMM Lifestyle Books are distributed in the UK by Grantham Book Service, Trent Road, Grantham, Lincolnshire, NG31 7XQ.

In North America, IMM Lifestyle Books are distributed by Fox Chapel Publishing Company, Inc., 1970 Broad Street, East Petersburg, PA 17520, www.FoxChapelPublishing.com

10 9 8 7 6 5 4 3 2 1

Illustrations by Dave Ansell

ISBN: 978 1 5048 0013 6

Printed in Singapore

Note: The authors and publishers have made every effort to ensure that the information given in this book is safe and accurate, but they cannot accept liability for any resulting injury or loss or damage to either property or person, whether direct or consequential and howsoever arising.